donor, once a very private matter (the use of a sperm donor has a negative emotional impact on infertile men), is now available online in a matter of weeks. Mr. Boni's obsession and diligence in searching for his mysterious biological father and his true origins kept me riveted as I empathized with the ups and downs of his journey."

—DAVID L. STERN, MBA, CEO, Boston IVF;
former global head, Fertility Franchise, Merck Serono

"Peter Boni takes us through his own comprehensive surprise journey of assisted reproductive technology, which uniquely sensitizes participants to the needs of those who science helps conceive. *Uprooted* unveils how scientific advances can cause harm, help, and happiness. The balance between these outcomes is an ethical fine line that he, with constructive compassion and factual examples, helps us walk."

—DR. DENNY SAKKAS, PhD, scientific director, Boston IVF

"*Uprooted* offers a personal view of the consequences of secrecy in the practice of donor conception. A prominent CEO, Peter Boni, uncovers a family secret that both validates and disrupts his life. Combining his extensive research of the early practice of donor insemination with his own genetic discoveries, Boni reveals the questions, uncertainties, and confusion he faced, as well as the exhilaration of connecting with genetic kin.

Boni provides a detailed history of the practice of donor conception, walking us through the social, legal, and religious climate that encouraged secrets in reproductive technology and stunted the advancement of our social narratives.

Secrets are no match for personal determination and the universal need for self-discovery, no matter what age. Boni doesn't stop his search until his parents' secret comes to light."

—JANA M. RUPNOW, LPC, author of *Three Makes Baby*

"*Uprooted* is a compelling mystery written from the perspective of someone who learned later in life that he was donor-conceived and is searching to find out the truth about his origins. I believe it will also be a valuable resource for those working in the assisted reproductive technology field."

—NAOMI R. CAHN, Justice Anthony M. Kennedy
Distinguished Professor of Law; Nancy L. Buc '69 Research
Professor in Democracy and Equity; director, Family Law Center,
University of Virginia School of Law

"*Uprooted* is a must-read for anyone looking to understand the donor-conception industry—yes, industry—and the effects it's having on people as they struggle to understand their origins. With the dawn of the mail-in DNA test, people's true identities are coming to light as quickly as opening an email. Deep secrets are crumbling as genetic bewilderment shakes the core of families from all walks of life. Peter J. Boni takes readers through his personal search for the truth about his own identity, while educating them about the donor-conception industry's checkered past. His deft writing manages to both entertain and educate. I highly recommend that anyone interested in the growing MPE ("misattributed parental event") phenomenon add this book to their library."

—EVE STURGES, MA, LMFT, therapist, writer,
and host of the podcast *Everything's Relative*

"*Uprooted* is a passionate telling of a secret that was intended to be taken to the grave . . . This is a story that not only provides a startling look at reproductive technologies but also serves as a wake-up call for an industry practice that no one talks about."

—*READERS' FAVORITE*

UPROOTED

UPROOTED

Family Trauma, Unknown Origins
and the Secretive History *of*
Artificial Insemination

PETER J. BONI

GREENLEAF
BOOK GROUP PRESS

The names and identifying characteristics of persons referenced in this book have been changed to protect their privacy.

Published by Greenleaf Book Group Press
Austin, Texas
www.gbgpress.com

Distributed by Greenleaf Book Group

For ordering information or special discounts for bulk purchases, please contact Greenleaf Book Group at PO Box 91869, Austin, TX 78709, 512.891.6100.

Design and composition by Greenleaf Book Group
Cover design by Greenleaf Book Group
Cover Image: Genome data pattern visualisation diagram, used under licence from Shutterstock.com
Author photo courtesy of Amy Rader.

Publisher's Cataloging-in-Publication data is available.

Print ISBN: 978-1-62634-907-0

eBook ISBN: 978-1-62634-908-7

Part of the Tree Neutral® program, which offsets the number of trees consumed in the production and printing of this book by taking proactive steps, such as planting trees in direct proportion to the number of trees used: www.treeneutral.com

TreeNeutral

Printed in the United States of America on acid-free paper

21 22 23 24 25 26 27   10 9 8 7 6 5 4 3 2 1

First Edition

To my family, old and new,
who unselfishly helped me discover
and embrace the truth!

"Three things that cannot be long hidden:
the sun, the moon, and the truth."

—Buddha

Contents

Author's Note

In these pages, I share what I have learned about the long, secretive, and sometimes scandalous history of artificial insemination. I do this using the backdrop of my own life's experience. I tell the story of my own donor-conception, which I did not discover until I was forty-nine years old, and show how this knowledge enabled me to better understand myself and grow as a person.

My parents had every intention of taking my clandestine beginning to their graves. The more I learned about the history and the sociology of artificial insemination, the more I discovered why they felt that way. I also discovered that I was not alone in having this secret kept from me. The number of donor-conceived people has grown to well over one million as advances in reproductive science have spawned a multi-billion-dollar assisted reproductive technology industry. The United States lags well behind other nations in developing and applying standard controls or regulations on this industry.

Amid a cadre of practitioners with high standards were a fraudulent and unethical few, along with a Wild West of gamete sourcing (sperm, egg, and embryo). Operating with little legislative accountability, its free market forces have contributed to the conception of dozens, sometimes more than one hundred, siblings from the same anonymous donor—siblings who may be unaware of their origins or of the existence of each other. The breeding of puppies enjoys greater oversight.

Perhaps more significantly, I was faced with what this has meant for society as a whole. With the advances of science, and the ease and accessibility of DNA services like ancestry.com and 23andme, secrets like mine have now been placed on open display. The former promises of eternal anonymity are now a heap of obsolescent assurances.

The donor-conceived ("semi-adopted") and the adopted share the same genealogical bewilderment: We do not know exactly where we came from. Added to those two categories are staggering numbers of people conceived in a Non-Parental Event (NPE) who are "misattributed," which means birth certificates may list the known parents as biological parents even if one or both are not, thus keeping the "artificial secret" alive. In recent years, social media support groups for the donor-conceived, adopted, and other misattributed children have cropped up, allowing these people, and their biological connections, to share and understand the implications and emotions of the experience from every conceivable angle.

Those who are aware that they are misattributed advocate to establish a "Bill of Rights" to abolish their parental anonymity, give them full and early disclosure of their genetics and medical history, put limits on the number of offspring per donor, identify sibling donor-offspring to one another, and define legal consequences for blatant fertility fraud. Society grows and changes, sometimes kicking and screaming and often well behind the advances of science, but it continues to grow. I hope that my story helps continue that growth. Truth emerges not only from a personal, genealogical history, but also from a scientific, legal, and sociological history. I began my journey by delving into the history of artificial insemination and assisted reproductive technology and then subsequently explored DNA analysis by two different vendors. Over the course of two decades, I discovered my own truth, which helped set me free.

PART I

THE DISCOVERY

Chapter 1

The side room in the funeral home was spookily dim on the warm, late afternoon spring day of my father's wake. The air seemed heavier than the purple velvet curtains that prevented the western sunshine from penetrating the window. As we stood there, the undertaker discreetly handed my mother a small, tightly rolled, dampened, sandwich-sized brown paper bag. He explained that it contained whatever my father had in his possession when his body was discovered.

As my mother unrolled the bag, the aroma of organic decay shot through my nostrils. Tucked inside was my father's thin, black leather wallet, now filmy and shriveled. Alongside it, wrapped in a torn white paper napkin, lay his bent and discolored 10-carat-gold wedding band.

When the wedding band fell into my mother's hand, my memory flashed back to the first time I noticed that ring on my father's finger. I was four years old and helping my dad dig holes in what was to be his garden at our new house. His hands were covered in earth and cow manure as he lifted a larger boulder from one of the holes.

"Watch this," he said, as he placed a small stone several inches from the boulder and held a piece of a sturdy limb atop the small stone. "The smaller rock is called a fulcrum. The stick is the lever. The longer the lever's handle, the less force we need to move this heavy boulder. Put one end under the boulder and the middle atop the little rock. Here, you do it and see." I can remember feeling that my dad was the most ingenious man on the planet as I watched that ring shine through the muck on what I believed were the world's strongest set of hands.

In the funeral home, my mother muffled her cry as she opened the billfold. It contained two waterlogged dollar bills, my father's driver's license, and a laminated family picture. I experienced an eerie feeling, and I am sure my mother felt the same thing.

He died with two bucks in his pocket?

"Let's wash our hands," she said. We never talked about the sad contents of my father's wallet, or their condition, again.

We did not talk about a lot during that time. Prior to his death, my father had not worked for over four years. We were too busy trying to survive and were just scraping by. My dad's illness consumed him and had an immeasurable impact on the two of us. His decision to take his own life added to that burden.

A loved one's suicide creates a wound that never heals.

In *Gone with the Wind*, Scarlett O'Hara looks toward the sky in the ruined post–Civil War fields of her family's southern plantation. She says, "As God is my witness, they're not going to lick me. I'm going to live through this and when it's over, I'll never be hungry again."

I experienced what I refer to as my "Scarlett O'Hara moment" as we buried my father in 1962. An indelible lesson in economics raced through my brain as I recalled the waterlogged bills that were all my dad had in his wallet when his body was found.

The family bank account was running on empty. My dad's once-pristine 1955 Buick Skylark, purchased with the proceeds from the sale of our suburban Boston house, which he had sold years before, had depreciated immediately after he drove it out of the dealership's lot. Now it was a seven-year-old rust bucket with a bad transmission and little value. He never owned a house again for the rest of his life.

I vowed to do whatever it took to get beyond that moment at my dad's funeral and to get out of the situation in which my mother and I had been left. In the aftermath of my father's suicide, I felt empty, needy, poor, abandoned, vulnerable, flawed, and alone. I developed a deep and fervent resolve to become successfully independent and financially secure. With a Scarlett O'Hara passion, I made a commitment to myself to never let my own future family feel that kind of emptiness. I would be affluent, recognized, respected, never vulnerable, and always strong.

I was just sixteen at the time of my father's death. Because of his mental illness, my mother and I already understood what it meant to make sacrifices. I spent my high school years without hot water in our home half the time. The kerosene stove in the living room provided heat and hot water to the apartment when it was lit, so there was no hot water in the late spring, summer, and early fall when it was not.

I was determined to avoid ending up like my father. I was not going to die poor, sick, and without having reached my full potential. I worked passionately and unconventionally to have a successful life.

More than three decades later, my life had changed considerably. I was living in an incredibly beautiful and luxurious neighborhood in the Back Bay of Boston, surrounded by the homes of nineteenth-century robber barons, just three blocks west of the Boston Public Gardens and one block south of the famous Charles River Esplanade. Life was good—stressful but rewarding.

I was a veteran high-tech CEO with a specialty for navigating companies out of hot water. I was the "Go-to Guy" for organizations in disrepair—I had developed a reputation as the "Go-to Guy" even before my father's death decades before. Then, it was about surviving; now, I was thriving. My pace was twenty-four/seven, with global travel, late nights, and constant demands.

Things were not all great, however. By this point in our lives, Susan (my wife) had reached the end of her rope with me. We were kids when we married, although we did not think so at the time. She had been my partner on a journey that had started when I received my draft notice the day after we met in a Boston nightclub. That draft notice meant that, in all likelihood, I would soon be en route to Vietnam.

The night we met, Susan gave me the telephone number to the apartment that she shared with a few other work/study college girls. I called her up at midnight the following evening. I was fully loaded with liquor, had my draft notice in hand, and was riddled with fear that I would lose the chance to ever be with this special woman. Even having just met her, I knew she was strong, resourceful, independent, ambitious, personable, and unbelievably cute. She had courage and a spirit of adventure that had prompted her to leave her small Pennsylvania town and move to a distant city without knowing a soul to further her education and improve her life.

Would I die or be maimed in combat? Would she still be available once my Army service ended? War has a way of accelerating human relationships. We crammed in a seven-month whirlwind courtship and hosted a cross-sabered military wedding while I was stationed at Fort Benning, Georgia.

At that time, I was an Army officer-cadet with barely $1,000 in savings. Now, we were living a senior-executive lifestyle. Some may see that as a prestigious lifestyle. My younger self certainly did, but that younger self had no clue, no role model for the personal and professional sacrifices it had taken to get there and stay there. Those sacrifices, along with some of the dark anxieties and fears I had dragged along with me from my past, had put a strain on my marriage.

My wife and I were struggling. My family found themselves

competing with my career for my time and attention. I camouflaged old emotional wounds from childhood and from my on-the-ground service in a jungle combat zone. All of these factors contributed to a stifled home environment with a shallow intimacy. I never discussed my fears, anxieties, or vulnerabilities. Wasn't that what my family needed? For me to be strong, accomplished, invincible?

Susan and I became estranged. I was living in Boston, and she was spending most of her time in our Cape Cod home. We decided to try to make it work and to start couples therapy. But before we began therapy together, I had agreed to start an eight-week counseling program with a therapist to address some residual issues I carried around about my dad, my upbringing, and my Army combat experience—all of which had impacted my behavior and my marriage.

During my first session, the therapist asked me to write a biography and describe my goals for therapy. I wrote about my father's mental illness and suicide. But I also wrote about all the hidden feelings of flawed inadequacy these issues had caused, and the bravado that veiled those issues. Of course, I shared that essay with Susan.

Despite knowing why I needed it, I had entered therapy half-heartedly. And I had selected a shrink who also seemed half-hearted about my healing. It appeared to me that he was completely enamored with treating a CEO of a global stock exchange–listed company with a high local profile. He even offered to play chess with me during our one-hour sessions instead of digging into my inner darkness. When he shared that his idea of success was to "move your wife back to the big bed," I bought it. Defining and resolving my issues were not important.

Around the time this was going on, my seventy-five-year-old mother required open-heart surgery. The surgery was a success, but she suffered a post-surgical stroke. Her survival was in question, but in the end she beat the odds. She regained consciousness three days later and entered rehab, determined to become fully functional once again. She gradually overcame her physical limitations and restored her memory. But not without some earth-shattering fallout.

The fifty-year-old locks on the iron gates that had guarded her untold secret disintegrated completely.

It was a seasonably chilly New England evening in late November 1995 when I got the big news that would change my self-perception and fundamentally alter my life. Snow flurries were in the forecast; my wife and I were having a low-key dinner at home after one of our therapy sessions when my business line rang. Global corporations run on multiple time zones. Troubled companies always need attention. A game-changing acquisition was due to be announced any day. "This could take a while," I told her as I ran down the hallway to answer the phone.

But the call was not about any deal. My therapist was on the line. "Your wife left me a message," he told me. "She has some information she said I needed to know in order to treat you. I haven't spoken to her yet. Why don't you?"

"That was a short call," Susan said as I returned to the dining room. "What's up?"

I refreshed her glass of the New Zealand Sauvignon Blanc we had opened to enjoy with dinner and poured a bit into my own glass. "That was my shrink. He told me you may have something to tell me." I felt agitated sitting there, thinking about this. "What the hell?" I snapped. "You're calling *my* therapist?" It came out as an unintended growl, but it was a growl, nonetheless. I was afraid, truth be told, that she had decided to throw in the towel and serve me with divorce papers.

We moved from the dining room table to the living room, she on the sofa and me on an adjacent chair. After a pause, she tearfully and softly said, "Your mother has been telling stories as she's been regaining her memory."

She took a sip of her wine. I took a glug of mine.

"She told me a new story," Susan continued. "And I don't even know if it's true." I could see she was choosing her words carefully. "With all that you have on your plate, I was going to wait till the weekend to tell you, but I thought that your therapist should know now."

I was not known for having patience, but I sat on my hands and silently waited for her story to unfold. "Well, here we are," I replied, feigning compassion for her tears.

"Your father wasn't your biological father."

In that moment, my life turned upside down. Everything I knew about myself, about where I came from, about who I was, changed in an instant. Everything I thought I knew was challenged with six simple words. I would go on to discover that my conception was not the product of an affair but had been achieved through a process to which my dad, the only father I ever knew, had fully consented.

Thus began my quest to finding my new identity. For the next twenty-plus years I was compelled to search for my biological father. That search for my genealogical roots wound up being multifaceted— it comprised my journey to know and understand myself in the face of this new information, research to understand the history of artificial insemination, and discovery of the evolution of social attitudes toward artificial insemination before, during, and after World War II (WWII), when I was conceived.

In the beginning, I used the old gumshoe tools of an investigator: interviews and library research. As science and technology advanced, I harnessed the power of research on this new thing called the *internet*, and then I added DNA analysis made available through the internet. As a result of these efforts, I began to uncover old secrets that would eventually make me feel whole.

Chapter 2

A Great Lakes–style winter thundered through Cape Cod like an enraged, snorting bull during the winter of 1961–1962, with record ice, snow, and frigid nor'easter winds that blasted into early April. Saltwater harbors had thick ice. Fishermen's boats were frozen to a standstill at their shallow water moorings. On my sixteenth birthday just before Christmas, adjoining rivers and freshwater streams were frozen solid.

Then, with what seemed like a flick of a switch, the sun broke through the Easter weekend clouds. A long-awaited spring was in the air like a fluffy blue blanket fresh out of the clothes dryer. April's warmth gradually freed the harbor's myriad frozen and captive commercial boats. Inland waterways and streams were at last flowing once again.

It had been a long and hostile winter. New England social conversation often revolved around the fickle weather. Cape Cod's economy ran on tourist dollars, fishing, and cranberry farming. Everyone was abuzz with the excitement of spring and the hope of a prosperous new season.

For me, my dearest hope was that my father would come back. He had vanished nearly a hundred days earlier, and I was not sure he was ever returning. Then again, in the condition he had been in when I had last seen him, I was not sure I wanted him to.

Don't misunderstand. I loved my father. When I was a young boy, I believed I had the best dad in the universe. We did everything together. We read the funny papers and the classics side by side. He took me to zoos and museums, encouraged me in my studies, and helped me with my math homework. He taught me to play cards, checkers, and chess; to fly a kite; to fish in freshwater ponds and ocean waters; to dive into waves at the beach; to ice skate and aim a hockey puck; to play baseball; and to throw and catch a football.

We viewed ourselves as visiting Bostonians when we went to see the Boston Red Sox and New York Yankees play the White Sox at Comiskey Park. Together, my dad and I saw some of the greats. We saw Ted Williams and Mickey Mantle hit home runs, Whitey Ford pitch as Yogi Berra caught, and Casey Stengel get tossed from a game for throwing a tantrum over an umpire's call. He also took me to historic Wrigley Field. We sat behind home plate to see the Chicago Cubs play two of the best teams at the time, the Brooklyn Dodgers and the New York Giants. We shared some magic baseball moments: Willie Mays shagged an incredible fly ball in center field, Sandy Koufax pitched while Roy Campanella caught, Jackie Robinson stole third base, and Ernie Banks scooped up what looked to be an unplayable hit and turned it into a double play. I was never short of adventures to tell my small town, stay-at-home paternal cousins.

My dad was affable, honest, kind, generous, and overly hospitable to everyone. He was his immigrant family's hero. He was the only one to leave provincial Cape Cod and venture into a bigger world—first to Boston, where I was born, and then to Chicago for ample work at higher wages, with a side trip to Florida for added measure. He made a living from defense contracts that needed his tool and die-making skills. A lax disciplinarian, he spoiled me, his only child. He was always loyal and worked hard for his family. He set a terrific example.

Except when he got sick.

My dad suffered from unipolar depression, which could be debilitating. When he was a younger man, he could shake off the occasional bouts of depression—until one day he couldn't. His depression came in waves and lasted several months. It became a parasite that burrowed deep within him.

While in the throes of depression, my father could not work. He just moped. Every once in a while, he attempted suicide. Once he put the gas stove on but did not light it. He tried overdosing on pills several times, but someone always intervened. As his depression grew, it led to periodic hospitalizations.

The absence of a regular paycheck soon depleted our savings. My mother waited tables for scant wages, but she could stretch a dollar further than anyone. She bought my clothes from the sales racks of second-hand stores, which was not an issue for me until I noticed girls. I knew I could not get anyone's attention or adoration in the clothes my mother bought for me, and I knew my mother could not afford to buy me more stylish and tasteful clothes. I had to find my own way to make money to purchase my own clothing, but was an eleven-year-old even employable? What kind of work could I possibly do?

One of the things my dad attempted to teach me when I was very young was woodworking. From his basement workroom which he set up with all kinds of tools, my father gracefully created useful things. I had little skill and even less interest in woodworking or the tools for it. Still, he persisted in trying to teach me. "You have to figure out how to use these things if you are ever going to make a living for yourself," he told me. Even at that young age, I knew I would never aspire to become a craftsman, but I had the aptitude to sell whatever anyone made. The idea that I could make money selling things brought me full circle to my father again.

Wherever we lived, whenever he could, my father always had a garden and always grew more than we needed. At six years old, I started collecting the overage, placing it into baskets, and ringing doorbells to sell Dad's beautiful vegetables to the neighbors at prices comparable to those at the local grocery. I intuitively understood that I did not have to offer lower prices than the store because not only were they better quality, but I was delivering them right to people's front doors.

A few years later, with that understanding as my business background, I invited a few friends on a bus to Maxwell Street, the center of Chicago's South Side wholesale district. With cash saved from birthday and Christmas gifts, I purchased an assortment of household and gardening supplies; seeds, potting soil, brooms and such. With my cadre of sixth-graders, we lugged home as much as we could carry. I repackaged the goods in my living room for resale, went out and sold them door-to-door, then returned to Maxwell Street and replenished my inventory. I felt empowered that I was on my way to earning my own keep.

Somehow, I finished the month with what my wholesale suppliers labeled "a genuine profit." I bought my father a new shaving kit and my mother a matching skirt and blouse. I selected an updated wardrobe for myself at Marshall Field's and had enough cash remaining to double my original savings.

I repeated this success, selling everything from Christmas cards, which were successful, to Valentines, which were not. But however much I earned, it was never enough to pay the rent.

My mother began to crack under the strain of trying to hold things together. When under stress, her decision-making became erratic and unsound. These poor decisions were about small and meaningless things at first; eventually, however, they affected the big things.

When she lacked the funds to pay the rent, we moved—five different times over the course of a year and a half. Each time we moved into a different school district in a lower-rent neighborhood, only to move again shortly afterward to another neighborhood a notch lower.

At one point, she believed that her sick, periodically hospitalized husband might be able to find less demanding work at the local university in Gainesville, Florida. A friend who worked there had beckoned him to give it a try. We moved into a decrepit trailer park. She hated the late summer heat, the humidity, and the insects. And my dad became even sicker. After a month in Florida, we trooped back to Chicago's lower-rent South Side. We moved three additional times over four months, from the end of August to late November of 1958, as I began the eighth grade. My dad found defense work, but with his recurring depression, he could not remain on the job and was hospitalized once again after another suicide attempt.

I made what I thought was the mistake of my life by running from a South Side bully who had threatened me with a knife on the school-yard. Like chickens, school-aged boys have a pecking order. The new kid in the neighborhood has to fight for his rights. Now every other kid in the chicken coop seemed to be testing me. Could I be preemptive and go after a bully on my terms?

From experience, even at such a young age, I had already learned that when you have no control over disruptive circumstances, you can still take control of how you react to something out of your control. I learned to control my reactions with purpose and direction. I was not about to engage in a knife fight, but I knew I could force an intelligent conversation with my overstressed mother who was distracted by her own struggle to survive.

At twelve years old, in our small, dreary third-floor apartment in the slums adjacent to the Chicago stockyards, where the outside air was thick with the smell of animal death, I decided to create a crisis to get my mother's attention. Amid our dwindling personal possessions remained my father's fishing tackle box. I stuck his old, rust-laden folding fishing knife in my pocket one day before school. At the breakfast table, after finishing a bowl of corn flakes, I made sure my mother saw the knife's handle protruding from my pocket as I rose.

Wearing her white uniform and preparing to go to work at the local deli after I departed for school, she turned as white as her uniform. "What's this about? You can't go to school with that!" she screamed. I could see the pupils of her hazel eyes dilate behind her black-framed glasses and the horror pour over her stress-lined face as her slender five-foot frame shook. I let her shriek linger. Only the ticktock of the kitchen clock filled the silence.

Then I spoke in a calm, adult manner. *Like my teacher,* I thought. I related the schoolyard incident and told her that I felt like she had been running from a house fire toward the first door that she found, even though that door was not leading to safety.

Then I made my calculated move. "How many times are we going to move, Mom? It's not even Thanksgiving and I'm on my third school this year. My ninth school since we left Boston. Have you been counting? And this one is in a dangerous slum. I need that knife to protect myself."

She sat silently as her eyes welled with tears. Once pretty, perky, brunette, and young, she looked fatigued, gray-lined, and old to me. She asked softly, with a defeatist sigh, "What can we do?"

There was my opening.

"We're alone here. Dad's in a hospital. We have no family support. Let's go back to Cape Cod where he has a family that adores him. Maybe I can have a life, too."

I clearly felt I was the adult in the room. I did not realize then that I would encounter other dysfunctional situations and that I would use the same preemptive sense of purpose in the face of disadvantaged circumstances. Throughout my life I would use bluff and bravado to reposition myself in a better place. And my mother seemed to need and welcome the direction.

I had given her a call to action. She did not take long. Within a week, we had discharged Dad from the hospital. My father wept as he got in the car, telling us, "I thought I'd be there forever."

More than a decade later, while engaging in a bullet-riddled helicopter extraction from a Vietnam combat zone on the edge of the Cambodian border, I silently flashed back to that very moment. I felt like I had saved him—that I had extracted him from a hostile combat zone and flown him to safety. It was bone-deep frightening and just as exhilarating, all at the same time.

We barely spoke as we drove back East over the course of two sun-filled days. Dad sat silently on the passenger side as Mom drove five hundred miles a day, with an overnight stay at a budget motel with two single beds plus a cot. I played "I spy," or read books in the back seat to pass the time.

Three winters' worth of salt on Chicago's streets had done a dam aging number on the rear fenders of their once-pristine car. Now it had fiberglass patches, brush-painted black, to cover the holes. That did not matter. To me, we were traveling in a black limousine to a better life . . . at my strategic direction. I felt empowered, with adult-like authority over my future. I could start and finish high school in the same place. My parents would have my father's family close by for emotional support through my father's episodes of sickness.

There was so much I did not understand.

We arrived on Cape Cod in time to have Thanksgiving dinner in 1958 with my dad's family. But the safety net was not a safe haven. The silent glares at the table felt like nails hammered into fingertips; there was a deep tension that would continue for the remaining three or so years of my dad's life.

Sure, his family loved him. But old-world Italian prejudices took over. The adults in the family spoke in whispers in adjacent rooms about his illness. Their attitude was that it reflected upon them, too. Their hero was flawed.

Keep it quiet. Don't talk about it. Don't let anyone know, or they will think we are flawed, too.

We rented a run-down four-room apartment where my dad spent his non-hospital days. It occupied a single-story wing of a two-story maroon-stained cedar-shingled rickety-old Dutch gambrel house with pre-WWII–era plumbing and fixtures. The structure sat diagonally across the street from a local strip mall, the Gateway Shopping Plaza. It was a fifteen-minute walk to the small center of town, a grocery store, and the hospital, and I could walk to school.

My mother and I both went to work. It was important to me to remain on the school honor roll. I wanted to buy my own clothing and fund my own social life. I wanted to go to college. Who could plan on a scholarship? I needed to fund my own education.

I withdrew from playing high school football and baseball so I could work after school and on weekends, either jerking sodas and scooping ice cream cones at Gateway Pharmacy's soda fountain across the street or stocking shelves, bagging groceries, and ringing the check-out register at the town's primary grocery store, the First National. Once I received my driver's license, I delivered flowers on Sunday mornings for Irene-the-Florist rather than going to church, and I served at catered events for the neighboring Mattapoisett Inn.

My mother worked a daily split shift in the local hospital's kitchen. Sometimes our dinner at home was whatever she brought home from that kitchen. We pooled our money, but it seemed like there was never enough. We lived paycheck to paycheck.

Over the next three years, my father's depression grew deeper and darker. While visiting my mother's brother in Boston, Dad, who was silent and sad throughout the visit, abruptly stood up, said good-bye to everyone, and walked outside into a cold, pouring December freezing rain. My Uncle Bill stopped him gently outside the front door. Then Bill and my mother drove my dad to nearby Massachusetts General Hospital. He was subsequently moved to the psychiatric unit at the state mental hospital in Taunton, about forty miles away. That was the first of several times that he would be admitted to the state mental hospital over the next few years.

"Taunton means mental head case," my dad's family whispered about his illness. They looked for scapegoats to explain away my father's increasingly deep depression. Their fingers pointed harshly, like deeply penetrating lasers. His sister accused my mother of committing my dad as a way of terminating her marriage. When I misbehaved in a typical adolescent manner, his adult nephew ambushed me at his front door. I was a "no-good bum of a son. It's no wonder your father is sick." I was labeled as flawed, too.

My father would occasionally spend a few weeks at home on arranged releases from Taunton, and when his depression became too debilitating, he would return to the hospital. *He shouldn't come home*, I often thought when his depression was so deep. During those times, at fourteen and fifteen years old, I would often scan the apartment fearing I would see signs of a murder/suicide when I returned home from a school basketball game, dance, or date.

In one last siege of depression at the end of October 1961, as US tanks and Soviet tanks faced off for sixteen hours at the border between East and West Berlin, my father held a kitchen knife to my mother's throat and threatened to end it for both of them. She stayed calm and convinced him to put the knife down, then bolted out of the house to call the local police. They arranged to transport him back to Taunton, where we visited him weekly.

Three months later, on a Sunday afternoon in mid-January 1962, we visited Dad in Taunton for his fifty-eighth birthday. I had turned sixteen a month earlier and had driven to and from the hospital. His once-bright crystal-blue eyes appeared filmy, watery, and lifeless. *The worst visit ever*, I thought. We had little eye contact and no conversation.

A quick good-bye led to my mother's rather perfunctory conversation with his doctor about giving shock treatment another try. The 1962 version of an electronic shock treatment was a traumatic, painful event, with the patient strapped to a surgical table and biting a stick. Was that something he would be willing to endure again?

We left the hospital and drove back home. We got a call at nine o'clock that evening. I answered the phone.

"Is your mother there?"

I recognized the voice. My father's doctor. "Mom, it's for you. The hospital."

We sat in the tiny, square linoleum-tiled kitchen, sitting across from one another at the speckled Formica table next to the pale green four-burner 1940s-era gas stove. I only heard her end of the conversation. She looked so fragile. "No, he's not. No, he didn't. I don't know. No note? Okay, let them know. Call me whether you find him or not. Let me know what's going on."

She hung up the phone and turned to me. "Your father isn't at the hospital," she reported. "They thought he might have come home with us. They don't know where he is. They're calling the police to file a missing person's report. We'll just have to wait and see. Maybe we'll do a search of our own." With a heavy sigh and a quiver, she added, "We need to get some sleep. We have a busy week ahead of us."

After that call, fourteen weeks went by without news. We tried looking for Dad ourselves, but had no luck. Friends and family called all the time. The questions were always the same. "Any news? Did they find him? Where do you think he is?"

They never dared ask the big question. But even then I was well aware that what they were all really asking was: "Is he alive?"

I didn't think he was.

The long-anticipated phone call came late on an April Saturday afternoon, just before Easter Sunday. I answered it, and as soon as the caller said he was from the Taunton Police Department, I knew what was about to unfold. My mother grabbed the receiver as she sat down in an armless black wooden folding chair behind my student desk. I stood behind her, my hands on her shoulders.

"Hello. Yes. Oh no!" There were sobs and anguish. "Are you sure? Do they have to? Yes, I understand. I'll let you know. Not what we had hoped for, but now we know. Thank you."

Through her tears, my mom explained what happened. "They discovered your father's body. The river's ice finally melted."

"He must have jumped from the bridge," I said.

I could picture him, walking jacketless from his hospital room after dinner to a small two-lane bridge that crossed the icy Taunton River. I could imagine what he was feeling. He was tired of being sick. Added to that was the dread of having more shock treatments. He had had enough.

There, he jumped. No note. No nothing. There was always a bit of a current under the center of that bridge, which kept a thin spot in the thickened ice. It had formed a predatory mouth that had swallowed him whole.

"It's been over three months. How do they know it's him?" I asked.

"Dental records match," my mother replied. She added, "Expect an article in tomorrow's *Standard Times*. I don't know how big. It's news."

"His family is going to love that," I said with a mournful sigh. And I also thought that my name would be on display for the world to see . . . his flawed son.

"We need to make funeral arrangements and let them know where to send him," she said. "Can you call the family?"

I had been mentally preparing for the worst news possible. A voice inside my head screamed, "It's my fault!" I silenced it. Instead, I told myself to be strong. I called my mother's family first.

They acted just as I had expected them to act, with love, sympathy, and support. "We're so sorry. What can we do?" they asked me. I was not weak or flawed in their eyes.

But in the collective eyes of my dad's family, I was the son of an embarrassment. I had to brace myself to update my father's family. I did not expect the process to be easy. They just could not reconcile their memories of my father with the deterioration of his larger-than-life persona. I mentally recapped all the family lore before making the calls.

My dad, named Nando, was the youngest of four children and the only son of Northern Italian immigrants who settled near paper and steel mills on Tremont, the west side of town, at the end of the nineteenth century. Industrial Revolution–era immigrants tended to settle alongside and support one another. Back in the day, Tremont was a Northern Italian neighborhood. The majority of its residents hailed from the Bolognese region. Nando's oldest sister, Emma, nineteen years his senior and the only sibling born in Italy, married a forty-year-old Italian immigrant when she was twenty. Her husband owned a meat market and grocery store that he co-founded with her father. They used a horse-drawn wagon to deliver food and goods to the surrounding community.

Nando and his other two sisters were raised in the same house-hold with Emma's four children. Biological aunts and an uncle were like older sisters and a big brother to Emma's four children. Frankie was Emma's oldest and was just two years younger than Nando. He eventually took over his father's store.

Nando had acquired near hero status with this group of siblings, nieces, nephews, neighbors, and friends. They reverently recited his history with folklore enthusiasm at many a family gathering. He had no shinbone in his scar-ridden left leg. At fourteen, he had severely gashed that leg in a bicycle accident. The wound got infected and would not heal. He was diagnosed with the bone disease osteomyelitis, which made his leg a candidate for amputation. On the day that World War I

(WWI) ended, November 11, 1918, the family admitted him to a specialty clinic and school for crippled children, located in the western end of Massachusetts, some 150 miles from Cape Cod.

Using experimental procedures, his doctors gradually grafted away his entire left tibia. They then fused his fibula to his kneecap joint. Throughout a twenty-four-month series of twenty-six painful operations, they administered shots of morphine each time to help him endure the pain and heal.

Fitted with a brace to enable him to walk, Nando made his way as both an accomplished student and highly motivated physical therapy patient. Five-foot-six, strong and muscular, with wavy brown hair, sky-blue eyes, and a ruddy complexion, he was a handsome young man. The school for crippled children he attended had, of all things, an intramural football program, to drive home the message "Don't let anything hold you back." Athletic and coordinated, my dad often recounted how he learned the game and became a quarterback. He had a flair for the game's strategy and knew how to call the plays. He could read a defense, handoff the ball, fake, and throw a precision-pass with a better-than-average chance of not being intercepted.

After two years of treatment and physical therapy, his brace came off. Could he walk? His left leg was an inch and a half shorter than his right. With his doctor's guidance, he bought a pair of elevator shoes and took the lift out of the right shoe to even out his legs. He walked, jogged, sprinted, and jumped with no limp. He was finally ready to return to his regular school and life.

My father's hometown had no football program, so he lobbied the school leaders and town administration to form a team. No deal—it was too frivolous and expensive. Not to be denied, Nando recruited his own team, comprising current high school athletes and recent graduates. He asked for volunteers to help him coach. Uniforms were makeshift. He set up a game schedule with other makeshift teams from neighboring towns.

Nando's team knew that they absolutely had to protect their quarterback's bum leg, but in the final play of the last game of the season, a tackler got through the line. Nando gave him a superb stiff arm that knocked out a couple of the burly would-be tackler's front teeth. His handoff to a big fullback resulted in the game-winning touchdown.

Fans of the opposing team ran onto the field to attack the quarterback, but his team furiously protected him. They won both the game and the fight after the game that day. The townspeople enthusiastically talked about the thrill of their local team's victory throughout the winter and into the spring. By the next fall, the town high school, feeling the pressure and encouraged by the success of a grassroots team, announced the start of their own football program. It would be led by the coaching staff of my dad's makeshift local team.

Before long, Dad started to get restless and longed for adventures that he knew his old Italian neighborhood was unlikely to deliver, so eventually he moved to Boston. He studied mechanical engineering and worked on the side as a skilled machinist, a tool and die-maker. But with his savings depleted and insufficient income to support himself and contribute to his aging parents' living expenses, he could no longer fund his engineering education. Rather than returning to Cape Cod, he moved to Detroit and gained practical mechanical experience and additional training with Ford Motor Company.

During the Great Depression, with much of his family and friendship circle unemployed and car-less, always-employed Nando returned to Cape Cod for visits in his rumble-seat Ford convertible. He would gather ten or more friends for top-down rides into town and silently picked up the tab for the party's food and drink.

In his early thirties, bachelor Nando introduced his teenaged fiancée to his circle. His best friend, Ray, recounted how Nando had done a double take upon meeting his attractive, effervescent brunette bride-to-be at a Boston house party two years prior.

He announced to Ray, "I'm going to marry that girl." She was known as "Chick," an affectionate nickname used by lifelong friends and family. Only her parents used her given name, Eleanor.

Ray told him, "For Chrissake, Nando, Chick is just sixteen. You're thirty-two. And she's not even Italian." Everyone within earshot laughed out loud.

My parents got married in a gala Boston wedding and shortly thereafter moved to Newport, Rhode Island, around the time of the outbreak of WWII. The Newport Torpedo Station had major quality problems. A faulty firing pin was deemed to be the cause of way too many torpedo duds being fired at enemy ships. Dad was the tool and die savior who crafted the fix. Mom and Dad loved Newport, the harbor, the boats, and the beaches.

Torpedoes were in low demand once the war ended. With my mom three months pregnant, my parents, along with their Welsh terrier, Rusty, moved from old-Yankee Newport to Italian immigrant–heavy East Boston. General Electric's neighboring submarine and jet engine plants needed tool and die experts. Rusty was rather cranky and snappy in a one-bedroom apartment in a traffic-heavy city over a variety store. Where were the wide-open spaces? Where was the beach? He was unhappier still once he shared that one-bedroom apartment with me, a new baby boy.

When my parents hosted a New Year's Eve party to introduce their new baby to their friends and family, Rusty chewed to smithereens all the fur and leather coats tossed on the bed in the apartment's only bedroom where he was held captive during the party. Mom recounted how she was beyond seething but used her good humor to get what she wanted. "We have a choice, Nando. Give the baby to Frankie and keep Rusty or give Rusty to Frankie and keep the baby."

My dad's brother-like nephew, Frankie, had a soft spot for me; we shared the same mid-December birthday. Frankie ended up taking the dog. Rusty became a well-fed fixture in Frankie's grocery store and meat market, where my dad made periodic visits to his gleeful dog.

By the end of his life, Rusty was deaf, crippled by arthritis, and blind, but his sense of smell was as keen as ever. He continued to make his morning rounds from the store to the nearby cranberry bogs and through the neighboring farm with the aid of that sniffer. Upon returning from Chicago, my mother took my dad to visit Frankie in his store and, of course, to see Rusty. With one sniff the old pooch became a jumping pup, so excited to get a pat from his old master. That scenario always raised Dad's spiraling spirits.

Then one day, Rusty did not return from one of his morning excursions. Six weeks later, once the ice melted on the flooded cranberry bog nearby and the water had been drained, Rusty's stiff and lifeless body appeared. He must have fallen through the thinning ice and drowned.

With all this family history swirling around in my head, I set myself to the task of notifying my father's family of his death. I thought, *Let's start with a friendly one.* When I told Frankie what happened, he choked back his grief. "Oh, no! Just like Rusty." I stifled my emotions. I knew I had increasingly difficult calls to make.

Next, I was going to speak to one of his sisters, one of my aunts, who had accused my mother of committing her brother to a mental hospital in order to free herself from an unwanted marriage. Now her brother had just killed himself. I was sure she would blame my mother. And she did.

One of Frankie's brothers had labeled me "a no-good bum of a son" who was responsible for his uncle's sickness. Now, his uncle was dead. The other brother was fuming that my mother was the reason his friends and work colleagues knew his uncle was mentally ill. Now that news would be in the Sunday newspaper. And it was—a front-page bulletin.

In any Italian neighborhood, Guidaboni was just a typical multi-syllabic name with lots of vowels. In his travels, however, Nando was conscious of early twentieth-century prejudice against Italians. Being laughingly called a "guido," had a negative connotation. No son of his was going to be called a "guido." Five weeks after my birth, Nando legally shortened the family's name to Boni. His middle name, Joseph, became his new legal first name, although to his family and friendship circle he was always Nando Guidaboni.

That Sunday, on the front page of the *Standard Times*, a small, thick black-line-framed box in the bottom center highlighted a late-breaking bulletin. It named a missing patient from the state mental hospital in Taunton whose body was discovered in the river as the ice melted—an apparent suicide. Nando's family and friendship circle knew him by his given name. The article used his newer legal name. The bulletin might as well have been invisible to that circle. Who was that? No one recognizable! Invisible for them, maybe, but not for me. It was a giant flaw for the world to see, right there on the front page!

Over the years, I had fabricated a story as a defense to cover my father's shameful illness. I had fibbed that this story had been told to me by my father's doctor. The twenty-six operations on his left leg and all the pain-killing morphine he had taken had altered his body chemistry, especially his brain chemistry. I had suggested to my mother a while back that this was the root of his depression. She bought it. I relayed it to my father's family and countless others over the years.

Thinking of my father in this way kept me protected. At the time I believed my dad was my flesh and blood, the tree from which I had sprung. I needed to believe he had not *inherited* his depression. I reasoned that if he could not have inherited it, then it was something that had happened to him. Therefore, there was no way I could inherit it from him. His fate would not be mine. Nothing in what I thought was my DNA could link me to his depression. Except, of course, for what it was doing to me during my formative years.

I rarely shared my painfully disruptive childhood and adolescence with anyone because I believed others would judge me. I thought they would abandon me just as my father's family had abandoned him. As I made new friends in college and in the workplace as I got older, I would dread getting asked, "How did your father die?" Only when I truly felt like I could trust that I would not be abandoned, which was seldom, would I take the risk and share some of the experiences about my dad, always adding this story I had fabricated.

Because of my father's depression and suicide, and the fact that I feared his mental illness could be passed down to me (despite the lie I told myself and others), I did all I could to prove to anyone, myself included, that despite my fear of being considered weak or flawed, I was strong. I was invulnerable. I was successful.

A day after my sixteenth birthday, just a month before my father disappeared from the hospital, I started on-the-road driver's training given by my high school. The instructor, retired coach Ray Ertle, was lean and muscular. He stood an imposing six-foot-three, with a young man's thick, full head of hair. To look at him, no one would have ever guessed that he was seventy-two. The only hint was his temporary slump as he emerged from behind the passenger-seat steering wheel and pedals of the training car after an hour's lesson.

It was snowing that December morning, and this was the first time I had driven in ice and snow. When my father was not hospitalized, he had taught me how to drive in our dilapidated Buick. Over two spring seasons, I had learned to maneuver that rust bucket on the dry, clear, lightly traveled roads going to and from the town dump. But now the roads had scarcely been plowed after an overnight blizzard, which had dumped a foot of heavy, wet snow.

"You have incredible luck," Mr. Ertle told me as we got in the driving school's brand new 1962 beige Ford Galaxie. He was as animated and enthusiastic as he was when coaching his championship football team in its season-defining game. "This is the most ice and snow that

I've seen in December on the Cape in over fifty years. It's going to be one bear of a winter. I'm going to teach you to be the best winter driver in the entire universe. For the rest of your life, you'll be the 'Go-to Guy' for anyone who needs a driver on hostile roads."

That struck a chord with me. Who would abandon the "Go-to Guy"? The Go-to Guy is skilled, a leader, held in the highest regard. I would find a way to always be that Go-to Guy.

With that fire consciously burning within me, I enthusiastically launched my adult life and opportunistically directed my career.

In 1995, on the brink of turning fifty, I sat in my Boston living room with my wife. I was an accomplished, grown man, the so-called Go-to Guy. I was about to learn that everything I had told myself—everything that had led me to become the grown man I was—was based on a secret I knew nothing about. Until that moment.

Susan explained: "Your parents went to a fertility specialist affiliated with Harvard Medical School. Your mom read about him in a newspaper after several years of trying to get pregnant. The doctor diagnosed your dad as sterile. He presented them with two choices: adoption or pregnancy via a sperm donor."

She paused to give me a moment to digest what she was telling me, as well as to regain her own composure. Susan spoke haltingly as her tears welled. She was clearly in a great deal of distress. I took two swallows of my wine and clutched the glass.

"You were conceived with the help of a sperm donor. It was all hush-hush. The donor was anonymous."

I felt a rush like a hurricane go through my soul.

The aftermath of a hurricane is wonderfully peaceful. The sky turns crystal blue. Enormous puffy clouds race through the clearing atmosphere in Indy 500 fashion. The humidity-filled air of the hurricane becomes purified, crisp, and clear.

I felt that post-hurricane rush as I recollected the sickness, the suicide, the feelings of inadequacy and abandonment, the need to be invulnerable and strong, the whispers that screamed at me, and the made-up story to hide my shame. I felt I was somehow now set free from him, unchained and unleashed to discover who and what I really was.

It was unbelievable, but very believable. The only grandmother I had ever known, my maternal grandmother, always referred to me as "the miracle child." I never asked why. The first grandchild is always considered "special."

Still. Why had I felt so different? Like I never quite fit in the family mold? My crystal-blue-eyed father was mechanically inclined. I had his blue eyes, but no mechanical interest or acumen. Did that gene just pass me by? And I was the only fair-complexioned blond child in the family. I passed my fair skin, tendency to sunburn easily, blue eyes, and blondish hair to my son, but he resembled no one we knew.

Inheritance is a complex formula.

Chapter 3

In spite of moving around all the time, it was important to me to excel in school. And it provided immediate status and recognition. "Hey, the new kid on the block is smart."

After high school, I attended the University of Massachusetts at Amherst on a partial scholarship. I majored in psychology for the wrong reason: to learn about my father's illness and its impact on me. However, I developed a passion for high-performance team dynamics and organizational psychology. I expected that I could use those skills in business after graduating. I would be the first in my working-class family to receive a college diploma.

In school, I did not talk about the skills of prioritization, focus, and goal-setting I had learned over the course of my life. I secured scholarships and balanced thirty-hour work weeks in between classes to pay for books and room and board, and to have spending money for dates. I mopped floors, waited tables, and tended bar with a fake ID. (To serve liquor, one had to be twenty-one.)

Then I started to make real money by selling Electrolux vacuum cleaners door to door. I offered that sweeper for free to an assistant engineering professor if he and his wife gave me introductions that produced ten new sales. Academic rivalry, I discovered, also made its way into their homes. Wives of full professors and department heads felt they deserved a better model than the entry-level unit, which was a deluxe sweeper with a power nozzle. My sales manager set up a promotion that I featured: buy the deluxe sweeper before Christmas break and get a free shampooer.

Between Thanksgiving and Christmas, I swept through the engineering department and broke a branch sales record. Electrolux recognized and rewarded that associate professor and his wife for their help. Their squeals of joy were hugely rewarding for me when I traded in their entry-level vacuum sweeper and carried to their home the deluxe version, with power nozzle and shampooer at no charge.

Since my Electrolux compensation plan increased its commissions as unit volume increased, my earnings for that month also broke a branch record; they covered my expenses for the remainder of my senior year and then some. My boss, Luke, gave me some advice: "You're very creative. Why not quit school and stay with Electrolux? You'll make a small fortune." I abruptly declined his offer. With the money I had left over, I bought a car to start my post-graduation life: a lightly used Buick Skylark convertible.

Boston was an exciting city in which to begin my post-college career in 1967. This was where my life started, where I belonged. All childhood stability had ended for me when we moved out of Boston and to Chicago in 1955. Now I was returning to stability. Except that in Boston, there were "silver spoons" (people from affluent backgrounds) everywhere. I was a working-class kid with a dead father. I felt embarrassed by my impoverished childhood in the presence of such affluence. Would my new peer group consider me from the wrong side of the tracks if I divulged that dysfunctional childhood and the entrepreneurial mindset I had developed to deal with that adversity? I did not talk about it.

The Boston workforce in the mid-1960s was brimming not only with silver spoons, but also with spoons carrying pedigrees—degrees from top-tier colleges and universities and the influential networks that went with them. The bankers, accountants, lawyers, venture capitalists, government officials, and healthcare professionals, not to mention many rank and file members in many respected organizations, wore their pedigrees on their sleeves like badges.

Social conversation during the workday was loaded with landmines: "This is where I went to school. How about you?" and "This is who I know. How about you?" But never was it: "This is what I have achieved."

Did they really look down on me? Or did I feel that way because I felt inadequate? I did not begrudge them their pedigrees, but I did not have their advantages. I needed to figure out a way to advance with a degree from the state university, without a pedigree.

Adding to the college experience was pressure of the wartime draft. If anyone dropped out of school to earn funds for the coming semester's tuition, he would lose his student deferment and be immediately drafted. I did not have powerful connections that would have allowed me to gain one of the scarce national defense deferments or even scarcer National Guard or Reservist posts. Gaining a deferment or making enough money to stay in school to keep a student deferment was literally a matter of life and death during the Vietnam era.

I graduated from college at the height of a wartime draft, which had exceeded 50,000 men conscripted per month throughout 1968. The US Army had redirected my own ambitions.

I took the officer's test and went to Infantry Officer Candidate School (OCS) in Fort Benning, Georgia, to gain a lieutenant's commission. About one hundred men out of my class of two hundred successfully completed the physically and mentally demanding program. My new proficiencies—weaponry, navigation though rough terrain, martial arts, parachute jumping, and leadership under duress—all provided new stature as an Army officer. I qualified, applied, and was accepted for advanced training at the US Army John F. Kennedy Special Warfare Center and School in Fort Bragg, North Carolina. Only those with what training officers called "Socratic Resolve" passed the challenging curriculum. I was highly motivated. Who could claim that an Army Special Operations officer was weak or flawed? Strong, resourceful, innovative, exceptional, and a leader, yes. But weak or flawed? Absolutely not!

My assignment to an elite force defined my collaborative sense of teamwork and clearly saved my life. I traveled alongside another elite unit within the 25th Infantry Division (CRIP, or Combined Reconnaissance Infantry Platoon) to win hearts and minds. After visiting remote villages deeply embedded in the jungle at the mouth of the Ho Chi Minh Trail for fifteen months, I lived to celebrate my twenty-fifth birthday with all of my arms and legs properly fastened, earned a few ribbons, and thankfully suffered minimum casualties.

The cultural climate in America at the time was tumultuous. Flag-burning anti-war rallies joined with peaceful protest marches; assassinations of political and social leaders were accompanied by both peaceful assembly and angry uprisings. The women's movement for equal rights joined the fray. Society had seemingly become unhinged during my time in uniform.

Five full decades later, pandemic-mania and left-wing/right-wing demonstrations, while disruptive, all seem mild in comparison to my 1960s recollections. And the absence of a military draft created seemingly benign resistance to a decade of Middle Eastern conflict.

Vietnam's anti-war movement and protests shared the daily headlines. A returning Vietnam-era soldier had negative status on the home front. Like many veterans, I experienced an ugly encounter with my fellow citizens while I was in full uniform upon my stateside re-entry, and my old baggage kicked in. Society had declared me abandoned, disregarded, disrespected, and rejected.

I buried the whole experience deep inside me and did not talk about it. Like most Vietnam War veterans at the time, I returned home, resumed my life, and remained silent . . . seething and angry, but silent still.

I took the decisiveness I learned while leading teams through the fear, stress, chaos, and confusion of hostile enemy fire into my post-Army civilian life. As a business professional, my career advanced as I leapfrogged over my peers. I took on risky assignments that many others shied away from. In special operations-fashion, I was righting tough situations where departments, divisions, and then companies had run aground. I set myself apart as the Go-to Guy.

I took on assignment after assignment, with each one proving to any onlooker, myself included, that I was strong, accomplished, respected; I was the Go-to Guy for high-tech organizations in disrepair. For sure, I encountered a number of setbacks that humbled me and challenged me to regroup from a momentary stall. But regroup I did, as the stress accumulated!

By the time my mother's secret came out, when I was nearly fifty, I had become the Go-to CEO for organizations in hot water. In the face of disruption and dysfunction, I shaped and led teams; together, we strategized, executed a game plan to deal with crises, and came out better positioned and more valuable. Reputed to be an inclusive leader with a well-established track record, I was certainly well compensated.

But now, with my origins in question, how many of those achievements did I owe to the experiences that had shaped my approach early on? The disruptive childhood, the state college education, the on-the-ground special operations combat? Or was there some unknown factor in my gene pool that had enabled me to become the person, and the leader, that I had become?

My experiences and how I dealt with them were mine. But what were the origins of my athleticism, my stamina and endurance, my intellect, and my tenacious will? Now, nearly half a century into my life, my origins were unknown! That troubled me. Mom had some answers, but her answers led to more unanswered questions.

Two months after my mother's surgery, I brought her home from rehab to a refrigerator full of snacks and pre-prepared food, along with schedules of visiting nurses, physical therapists, and Meals On Wheels. I gave her a week to readjust to being home before confronting her about the story she had leaked to my wife and friends about my conception.

Susan and I made a late afternoon Sunday dinner in early December and sat with Mom around her narrow oval oak dining room table.

"This is the first time I've been dressed up since my surgery," she joked. She wore a green skirt and flowered blouse instead of a hospital gown, pajamas, or sweatpants.

Once the food was consumed and the small talk tapered off, it was time to get to the heart of things. "Mom, we need to talk." I asked her about the stories she had been telling others while in rehab.

"No, that's not so," she quickly stated.

Susan was gentle but matter-of-fact and told my mother that she had, in fact, told the same tale of my conception to several visitors, herself included, while regaining her memory and functions at the rehabilitation facility. "It's time to put it out there," she added.

"I must have been delusional," my mother exclaimed in a panicked tone and with a nervous laugh. I recognized that panicked look—it was the same look I had seen in the slum kitchen in Chicago. But this time I had a different kind of knife in my pocket.

"Mom, I really need to know all of this. I've been carrying loads of baggage and fear about an inherited condition," I said. Susan sat next to me and stroked my arm.

Once again, my mother was quick to pounce. "His doctor told you long ago that it's not inheritable!"

She continued to deny everything until I told her that I had fabricated the story about my father's Taunton doctor telling me that his depression stemmed from altered brain chemistry due to childhood surgical trauma. Sobs of relief from a secret she had held tightly for fifty years erupted. Her tears overflowed and the truth came out. Yes, my father had been sterile. Yes, they had undergone a procedure to conceive a child the only way they could.

She abruptly lifted her frail frame from the dining room chair with the aid of her rehab-provided walker and fled to her bedroom down the hall. She cried. Loud, hard, and long. I had not heard her gasping sobs like that since the Taunton police sergeant called about the discovery of my dad's body thirty-three years earlier.

She reappeared in the dining room ten minutes later, blowing her nose. Her eyes were red and swollen. "I've held that in for fifty years. Aren't you mad?"

I had been angry when I was younger, but the anger I had felt was aimed at all the disruptive moving to different schools that I had experienced years ago. I had resolved that anger. My mother was in a tough spot and doing the best that she could.

I answered as tenderly as I could. "Why would I be mad? Because I was loved and wanted so much? That I obviously have a good gene pool? Sure, it would have been nice to have known sooner, but I'm just happy that I know."

I needed more time to come to grips with the new emotions I was experiencing. Who was this donor? How had living this lie shaped my persona? Did I shape what I had become because of some unknown biological factor, or was it how I handled the dysfunctional childhood that framed who and what I had become?

But in that moment, with my mother, I understood that this was not the time to put my questions on the discussion table.

"He was the proudest father imaginable," my mother pronounced. "You were his son, *our* son. He was your father in every way that meant something. He just wasn't biological. He'd be so proud of the man you have become."

I had a vivid flashback to my childhood. My Little League coach was short-handed. Dad had attended my every game. The coach asked him to fill in and coach third base. He reluctantly agreed: "Just this time." Dad was a shy guy. Coaching third base was not his style.

As a right-handed batter, I hit a line drive just over first base. It was not on purpose; it was due to a late swing. The ball landed in safe territory in no-man's land along the right-field line.

As I raced around the bases and approached third base, I heard him yell, "Slide!" and I dove headfirst. The ball was overthrown. Remember, this was Little League. He then yelled, "Run home!"

I scored across home plate and looked in admiration down the third-base line at him. In that moment, I saw the most delighted father. Delighted for me, no question! I agreed wholeheartedly: He was my dad, and I knew he was proud of me. But my love for my dad and my need to know more about my origins were not mutually exclusive. I had room for both emotions: I could love my dad and also need to know my genealogy.

Over the course of several days, my mother recounted all the details she could remember. She had read in a newspaper an ad, or was it an article, about Dr. Sims, a Harvard Medical School professor, with his office at 10 Beacon Street in Boston. Their first appointment was in the fall of 1944. In what paper or what magazine she had read this, she did not recall. Except for her husband and the fertility doctor, only her mother had known.

"The doctor first attempted to boost your father's sperm with injections. That didn't work. Then he extracted it, treated it in his laboratory, and inserted it in me. When that didn't work, he told us your father was sterile. We could adopt a child or do this fairly new procedure involving a sperm donor with a 10 to 15 percent success rate.

He would likely be associated with Harvard Medical School in some way, presumably as a medical student. He said to trust him to find a good match."

Just a week after having "the talk" with my mother, in mid-December 1995, my wife arranged a private dining room for a small fiftieth birthday dinner party at an upscale Boston restaurant with three of my closest friends and their spouses. We had earlier confided in them about our marital discord, separation, and therapy; they were all fully aware of my disruptive childhood and my father's suicide. There are no friends like old friends. They know and love you for who you are. At this dinner, I filled them in on my discovery.

Frank and Diane were a couple that we had become friends with in our old suburban neighborhood. We had known them more than twenty years. Our children were close in age. Frank and I co-coached a Little League team. Frank's son and mine were close friends. They were immensely relieved when I spilled the beans of my new discovery.

"Your mother started telling us this story when we visited her at the rehab center," Frank told me.

"Unbelievable, but so believable," I said.

Diane added, "We told her, 'You must tell your son.'"

Frank said, "She told us, 'I can't. He'll be so mad.' Good for you. Now, we're off the hook," he joked.

Ken, an environmental engineer, had been my closest school friend since we had moved back to Cape Cod from Chicago. He had an insider's view of my father's time spent in the Taunton mental hospital, his disappearance, and the reaction of his family. He was always there for me—supportive and never judgmental. Ken commented on how progressive my parents were to be "at the cusp" of reproductive science.

Mike, my college roommate, knew where all my skeletons were buried. We shared not only college shenanigans but also intimate peeks into each other's personal experiences. We had a brotherhood bond of trust forged over the four years that we had been roommates. That bond continued to grow over the decades after graduation. Mike knew me better than I knew myself. As a best friend, he could tell me the ugly truth with no flowers.

Initially, Mike acted surprised. "You never told me!"

I quickly clarified, "How about I just learned this and I'm telling you now?"

Over the course of our dinner conversation, he commented, "I know you. If it takes you twenty years, you'll scratch at every post and uncover the source of your biological roots—your donor."

As we were leaving for the evening, he pulled me aside and said softly, "You know, this therapy thing is a great idea for you. You've been packing heavy baggage that hasn't really been yours to carry."

I agreed. The secrets, both real and imagined, had exacted a heavy cost, paid in part by my wife and children.

No more secrets, I thought.

Next on the list of people to share my gift of knowledge with were my children: my son, Stephen, who at the time was a graduate student, and my daughter, Tracy, a college senior. They had just returned home for their holiday break.

We spent holidays at our Cape Cod retreat. It was a mild pre-Christmas week, 44 degrees, light winds, bright sunshine. Our children were not blind to the scars I carried from childhood, or the stifled post-traumatic stress disorder (PTSD) of a returning Vietnam combat veteran. Little by little as they were growing up, I had told them stories about the lessons of my own childhood and my Army experiences. Now I needed to let them in on their DNA; not what I *was*, since I did not yet know, but rather what I *was not*.

"Let's take a walk," I suggested to them. I wanted us to share some time alone without the distractions of television, phone calls, or possible holiday visitors showing up at the front door.

We all loved the Knob, a preserved wooded pathway around Quissett Harbor on the western edge of the Cape Cod peninsula. That hidden pathway opened up to a jetty from which you could see for fifteen miles across Buzzards Bay. The forty-mile-wide blue western vista of the southeastern mainland of Massachusetts was commandingly clear. In the north part of our view was my old high school home, and to the south was the historic whaling city of New Bedford. The view extended southwest to reveal the entirety of the Elizabeth Islands chain. The last island, Cuttyhunk, a frequent sailing destination of ours, glowed regally in the late afternoon sun. I intended to figuratively take them down a hidden path and put what I had learned in the open.

They told me they had commiserated together beforehand, "What's up with Dad?"

Their grandmother had recently had open-heart surgery and a stroke. Perhaps they wondered whether she was dying? We kept it no secret that their mother and I had separated for a time and were working through our marital issues in therapy. Maybe they wondered whether their parents were going to permanently separate? As we walked the wooded path toward a rock jetty on the ocean's edge, we all marveled on how stunning nature's beauty was around us.

With that as the backdrop, I unfolded their grandmother's secret, along with my feelings that had been pent up since childhood: the fear of abandonment, the fear of being perceived as weak, inadequate, and flawed. I had issues.

How about the kids? Did my parenting create issues for them? Was I too tough of a dad? I had needed my dad to be strong. But he was not. He became weak and sick. I wanted them to have what I did not have—a dad who was indestructible, with means. Was I too strong? Did the focus on my career success impact their well-being?

They were each about to embark upon their adulthood. Stephen was struggling to find a place in the real world for his creative passion. Tracy would soon graduate and begin her own journey. Would my baggage negatively impact their coming adult ride?

My son spoke first. "Dad, we had friends with parents who never traveled, but they were emotionally absent and unengaged with their kids. You were always engaged and emotionally supportive. If you're feeling guilty, get over it. We're fine."

When I shared that, along with beginning marital therapy with their mother, I had entered individual treatment with a new therapist, my daughter commented, "Dad, this is what I've always wanted for you—to heal."

We all huddled for group hugs and tears.

My mother's fiercely guarded fifty-year-old secret was a secret no longer. Not from me and not from the people I loved and trusted. "No more secrets," I repeated to myself.

After informing my children of the unique details of my conception, I spent the next few months digesting the information myself and asking my mother whatever questions came to mind. Given the post-acquisition step-up in my business activity, I had scant time for more.

Chapter 4

I was in the second year of a tough turnaround: to upgrade a once high-flying software company's aging technology, hold on to its key employees and customers, turn a profit, and increase the value of the company's downtrodden stock. It had caved from its rich IPO price to just under $1 per share when I walked in the door. The business turned a major corner once some newly introduced products gained traction. Analysts gave the stock a coveted buy recommendation for investors with risk tolerance. We had just announced a game-changing acquisition.

How would I cram in time to begin post-holiday therapy with a new psychologist? My new doctor specialized in high-achieving people who had never properly dealt with their trauma. In my case, the genetic identity trauma triggered flashbacks of a dysfunctional childhood, thirty-four-year-old grief for my dead dad that I had never allowed myself to experience, and the PTSD of war. He was deadpan serious when he said, "You hit a trifecta. Newly experienced trauma often resurfaces others that were long past."

Given my baggage, I had no compassion for weak, needy people who required a shrink. The social context in which I had grown up and spent my adulthood told me that weak and needy people were unsuitable for command. I had witnessed that attitude in both an Army war room and a corporate board room. Weak and needy people were inadequate and not up to the task. Not me! I kept all my perceived deficiencies disguised and under wraps. I had dismissed the therapy notion when Susan had suggested it for us ten years prior.

It was a disturbing reality that I needed a therapist's help, but I did. I thought I understood my personality. Years before, I had taken a Myers-Briggs personality test that scored psychologist Carl Jung's sixteen personality attributes. My results indicated the leadership profile, ENTJ: Extroverted, Intuitive, Thinking, Judging. This profile was rather rare—only an estimated 2 percent of the population has this profile.

I was a planner, but I was flexible enough to rapidly adapt to changing conditions. I was able to make tough, logical decisions without letting my feelings get in the way. And I had enough collaborative charisma to get people to follow me up a hill. Was I born that way? Did my experiences shape me? Likely, it was a combination of both.

That personality profile worked in the jungle as a Special Operations Team Leader and certainly in my chosen career. It was not working in my adult home. I had learned to deny my feelings. To top it off, my wife found that the privacy with which I carried my feelings, my invulnerable air, robbed her of an intimacy with me that she craved. For her, our relationship had not grown. It was shallow and incomplete. The leftover anger from a traumatic, life-altering Vietnam combat experience wore thin, too.

I was angry over the disruption in my life stemming from the lies of a US president; I was angry over how I was treated by my fellow citizens once I returned to the home front. There were other ways I might have built my character without going to war. I never talked about it. None of us vets did. I had enough flaws to camouflage. Why add "baby killer" into the mix?

In marital therapy, it was evident to us both that Susan carried emotional scars of her own from the sudden death of her father when she was thirteen—her own fear of abandonment. Our sessions put our mutual fear into clear context. We both worked to understand and better communicate our own needs to one another. We were getting somewhere, and we both wanted to go there.

I started to take my own therapy more seriously as I worked to save my marriage. I liked the Rogerian style of my new therapist, who was also a Boston University professor. He would ask me a question and repeat my answer so I could dig deeper into my answer. Then he was instructionally interpretive, treating me like both a patient and a student. That worked for me. He observed that I had built up a brick wall as deep as a city block over four decades of constructing my defenses. I did not need to blow it up, but perhaps I could gain some clarity by looking at my past in the third person.

"Construct a story in your head about someone else, who happens to be you. Then let's penetrate that wall, talk about your fear of abandonment and several decades' worth of post-traumatic stress accumulated between your disruptive childhood and your combat experiences. They've had their impact on your persona and your behavior." He added, "We won't eliminate the baggage that contributed to making you the person that you have become. We're all flawed. That's the human condition. But, in the big scheme, you're awesome. What we'll do is construct a handle so you can better carry that baggage."

I intuitively understood where to position the fulcrum. Without hesitation, I said, "Let's talk about all the whispers; the whispers that scream."

I had returned to school after my dad's 1962 low-profile funeral, imagining all the eyes landing on me as I opened the outside door and entered the noisy, busy pre-opening-bell hallway. Whispers reverberated: "Flawed!" and the sentiments of my dad's adult nephew, "You no-good bum of a son. If you had more to offer, he'd still be alive." I felt the piercing glare of their eyeballs and heard the deafening whispers throughout a very long day. "It's inherited, you know." At last, the final bell rang. I raced out for my afternoon job at the First National grocery store.

However, standing at the side door as I was trying to make my escape was the head of driver education, Ray Ertle. He tugged my arm, guided me into an empty adjacent classroom, and closed the door. He said gently, "I didn't realize you were Nando's son. You've got half a name."

I explained his name change.

"I'm really sorry to hear about your dad. If it weren't for him, I never would have coached at this school."

"How so?" I asked.

Mr. Ertle told me he had served as a volunteer coach on Dad's ragtag football team in the early 1920s. "He taught me loads about the impact of affection, caring, and taking responsibility for winning through teamwork. His poorly practiced players beat a superior team by protecting the daylights out of your father's leg. The town was so jubilant that they hired me to coach their new team. That was forty years ago." Then he finished, "Your dad was a quiet, inspiring leader. He led by example. Remember him for who he was."

He didn't say it, but I imagined the ending to that sentence: "Not by the sick man that he had become."

I had not shed one tear since receiving the phone call from either my father's Taunton doctor or the Taunton police sergeant. I had resolved to remain strong. But with Ray Ertle's empathy, I exploded in overdue sobs. The graceful and athletic arms of the aged driving instructor wrapped around me like the huge, nurturing limbs of a leaning, lumbering oak tree, loaded with leaves to compassionately shelter me and hide my crying from view.

Once I regained my composure, I said, "Thank you so much, Mr. Ertle. You have no idea how much I needed to hear that."

"Sure, son. Good luck," he replied.

I entered the First National with red and swollen eyes. The store manager greeted me with his own red and swollen eyes. "Someone else suffering from spring grass allergies along with me, I see."

Autumn mold pollen was my issue, not spring grass, but I replied, "Yeah, my mother has it, too."

My therapist and I spent a few months taking those whispers apart and airing them out to dry.

The end of winter in 1995 gave way to the beginning of spring in 1996. My big deal acquisition was in full swing. The integration of operations and Wall Street communications was intensely time consuming. I had grown enough to admit that I needed a break.

My wife and I spent that spring solstice weekend at our Cape Cod home and decompressed. March winds howled over the abutting marsh and adjoining Buzzards Bay waters from the southwest at 35 knots, with gusts to 50 knots. Sailors called it "a big blow."

I had moved some summer items from their winter storage in our basement to our garage in an early preparation for the coming beach season but had failed to fasten the basement bulkhead door. While asleep, I heard the roar of a gust and the *sproing* sound of the basement's bulkhead door being uplifted and heaved open by the big blow.

I was half asleep when I vividly saw my father emerge from the basement. Dressed in chinos and a white tee-shirt, he climbed up the steps, walked into the backyard, looked around, and admired the view.

Then I awoke, opened the window blinds, and peered outside. Four- to five-foot whitecaps from the wind-blown, churning sea rushed across

Buzzards Bay from left to right as the dewy morning light emerged. I was fully aware that I had been dreaming and would not discover my father standing on the back lawn outside. I looked for him, nonetheless.

But the message I had perceived was abundantly clear. My therapy was making progress; at fifty years old, I was no longer hiding my father. I was releasing him.

I wanted to extend this newly uncovered knowledge to more insiders within my family and friendship circle. Next on the "share-the-knowledge" list was my Italian cousin, and now pharmacist, Eddie. We had maintained our lifelong friendship well after his father ambushed me with his derogatory accusation. Eddie had his own issues with his father. We had a common bond, maybe even a common enemy for a time. But Eddie loved his father like a son should, and he was loved in return. His relationship with his father mellowed and grew as they both matured.

My cousin and I shared a deep love of the ocean and sailing. I invited him for an early season sail one June afternoon and offered him the helm. In the cockpit of my sailboat while on a starboard tack with modest 10 knot winds on a beautifully sunny 70-degree pre-summer day, I detailed the blow-by-blow of the story that had unfolded over the previous six months.

Eddie also had an insider's view of my childhood. As a child, he had spent time observing the family's adults whisper about my father's illness in the adjacent rooms. He had tried to protect me, to warn me to avoid his father, who was "on the warpath" looking for me, the misbehaving bum who he had decided was responsible for a mental illness in the family.

"How does this make you feel?" Eddie asked out of concern for me. "*Feel?*"

I never discussed my feelings. This was new for me. I had practice protecting myself. Revealing myself was something I was learning to do with the help of a therapist. Now I was doing so with my wife and my children—and now with my cousin who had a long-standing interest in my well-being.

"Both un-whole and free," I replied. "Un-whole because I don't know the whole story of my ethnic heritage; I may never know. And free, because the big inherited possibility that I thought I carried isn't really there. However, it has clearly shaped my behavior. That, in itself, has been my biggest issue. I'm just figuring out how to better carry my baggage."

We were both keenly aware that we shared common experiences, common family stories and traditions, common foods, and a common definition of hospitality—all part of our common upbringing.

In the absence of a common biology, Eddie coined me a "logical Italian," his "logical cousin." Now, this logical cousin needed to know his genetic origin.

Over the course of individual therapy, my doctor-professor and I discussed a lesson I had recalled from one of my psychology classes.

In the mid 1960s, a pair of psychologists had researched and treated troubled adoptees who sought the missing pieces of their identity. The duo feared that the absence of genetic knowledge impacted an individual's development and his or her feelings of belonging. The same could be theorized for donor-conceived children who possess knowledge of their maternal heritage only.

I mentioned this to my professor-therapist. He immediately responded, "Oh, yes, Erich Wellisch and H.J. Sants. I know their work well."

While I did not grow up with what those research psychologists labeled "genealogical bewilderment," as a fifty-year-old adult who had had the proverbial rug pulled out from under his genetic identity, I wanted more answers to my paternal genetics beyond being a "logical Italian." What about any implications to my health, the health of my children, or the health of my grandchildren?

For my entire adult life, the questions asked by doctors always added to my feelings of being flawed. They would ask about the illnesses in my family: Was there cancer, heart disease, or diabetes?

And they would always ask: "How did your father die?"

The answer, "suicide," always caused a wrinkled brow and additional questions.

"Do you ever feel depressed? Do you ever want to hurt yourself?"

No! Never! I am strong, invulnerable, a leader. The Go-to Guy.

When I first learned that my father's biology was not my biology after all, I was relieved. But as time went on, the desire to know from whence I came grew from an intellectual curiosity to a deep, burning need.

With that fire, I embarked upon my research. What were my paternal genetics? They were certainly different than I had thought. Would the answer to that unknown lead to a disease-ridden horse thief who was hung in the town square? I was preparing myself for truth. No more secrets!

Chapter 5

My mother had given me several clues, pieces of information to research that went back fifty years. Dr. Sims, his Boston office at 10 Beacon Street, his Harvard Medical School affiliation, and the 1944 newspaper ad. Or was it an article?

The internet was in its relative infancy in 1996. Nothing germane surfaced when I typed into a Netscape browser the doctor's name, "Sims," or "Harvard Medical School-fertility" or "1944 fertility clinics-Boston."

The Boston Public Library was just a few blocks from my Boston condominium. My daughter, an adept researcher and highly skilled sleuth, came home for spring break and joined me for our first Boston Public Library research visit.

Tracy shared my curiosity, perhaps even my need, to know my genetic origin: *our* genetic origin. Not only did she enjoy weaving together a good story, she was nurturing her dad and his need to know about his biological father. I loved that we had embarked upon this adventure together.

And I needed her help. The library's research assistant warned us that we would find little documentation from that era. Artificial insemination was a taboo topic in the press until recent times. Discouragingly, we uncovered nothing pertinent, but we both continued our individual research off and on for much of the remainder of the decade.

We started by inspecting 1944 and 1945 Boston telephone directories, both the white and yellow pages. We found no Dr. Sims in Boston. Dr. Simon, a family practitioner, did have a Boston office within 10 Beacon Street. So did a variety of specialists, insurance agents, stockbrokers, dentists, and an oral surgeon. Our search through microfilm of old Boston newspapers from the summer and fall of 1944 produced no fertility article or advertisement. And we uncovered zero information about Boston fertility clinics in general that existed in 1944. We did find information about sperm banks and fertility in a piece from the waning period of the Vietnam War era, not the WWII era.

Tracy identified the next place we could search: "Harvard Medical School has its own world-class medical library."

During several visits to neighboring Cambridge throughout the spring and summer of 1996, I focused on the name "Dr. Sims." There was no record of a Dr. Sims on staff at Harvard in 1944 or 1945. The only mention I found of a Dr. Sims was the nineteenth-century surgeon, James Marion Sims, reputed to be the "father of modern gynecology." Originally from a wealthy family of enslavers in North Carolina, he resettled in New York City. He was quite a historical figure, but he was not located at Harvard University in Cambridge, Massachusetts. He died in 1883. "Not even a close match," I concluded.

I questioned my mother again and again.

"His name was Sims," she repeated. "Your father would take a mid-week day off, and we'd either drive or travel on the train from our home in Newport, Rhode Island, to Boston for an appointment. We would stay with my mother in her Boston apartment the night before and head to the doctor's Boston office on 10 Beacon Street the next day. We used the stairs. It was on the third-floor rear. Then we'd return to Newport. That was over fifty years ago, but I remember it clearly."

"How about some other names," Tracy suggested to me. "Gram has never been good at names."

Simmons, Simpson, Stenson, Simon? Whatever the spelling of Sims or the derivative "S" name that we had concocted, our search always encountered the same dead end.

My search using my mother's clues to my origin all led to stone walls at both the Boston Public Library and Harvard's Medical School Library.

I turned increasingly to developing internet technology—first to research the history of artificial insemination and then to research the WWII sociological climate in which my parents made my conception decision and fervently held it as a closely guarded secret for so long. I likened myself to a dissertation-bound part-time PhD student and hoped that this would unlock additional genetic clues along the way. The entire research process captured bursts of my energy, in between sixty-hour work weeks, traveling 100,000 miles per year for business, and living life in between.

Tracy unselfishly volunteered as my research partner from the very beginning of my quest. She had been a college senior working deliberately toward her own future when her grandmother's secret unfolded. Two decades later, how she balanced her full-time employment, marriage, and motherhood of two young boys to effectively continue to uncover more details of my paternal connection is a wonder in itself.

Throughout two decades, I shared my evolving discoveries with my inner circle of close friends and family with a great deal of enthusiasm. "Hey, you'll never guess what I just learned!"

PART II

THE EDUCATION

Chapter 6

Tracy graduated from a top twenty university with honors. I was a proud father as she embarked upon her adult life. She used her limited spare time to continue her research on my behalf and discuss her findings with me, which were just as scant as mine. As yet, we had found no trace of a Dr. Sims from Harvard Medical School or any fertility practice at 10 Beacon Street in Boston.

At her suggestion, I spent time in Harvard Medical School's library and skimmed a 420-page book, written by Dr. A. Schellen and published in 1957, titled *Artificial Insemination in the Human*. While far more scientific than sociological, it was eye-opening. As it turns out, artificial insemination was not some new medical innovation that emerged during the twentieth century. To my surprise, it had a much older and rather checkered history.

Like much of medical science, artificial insemination emerged from a reproductive practice first used on plants and animals. I learned that for millennia, farmers in a host of societies practiced selective breeding to enhance their crops and better protect them from the ravages of insects, disease, floods, and drought. Stronger seeds that could better accommodate predictable harshness dealt by Mother Nature all led to richer, more bountiful harvests. It took several centuries to move the selective breeding practices from agriculture to livestock and poultry.

Medical advances throughout history have generally originated from experimentation on animals such as monkeys, mice, and dogs. Such was also true with artificial insemination, which initially focused on the selective breeding of equines and canines. And let's not forget the profit incentive for their fine breeding. In 1322, equipped with a clever early version of a turkey baster, or so the story is told, an Arab chieftain "stole" semen from his adversary's strongest stallion to inseminate his prize mare. She bore a "wonder" foal. Whether fact or tall tale, the Arabs were the first recorded people to use artificial insemination on their mares. It is no wonder that Arabian thoroughbreds are recognized as among the finest horses in the world. Seven centuries of artificial insemination has led to an excellent breed of horses.

Artificial insemination research moved in slow motion for more than two centuries until Dutch scientist Antonie Philips van Leeuwenhoek, renowned as "the father of microbiology," discovered sperm cells in 1677. Armed with silver- and copper-framed microscopes with lenses of his own making (some 500 lens designs in all, with the capacity of 275× to 500× magnification), van Leeuwenhoek is credited with not only the discovery of sperm cells, but also with discovering single-cell and multicellular micro-organisms, muscle fiber, bacteria, red blood cells, and blood flow in capillaries. Although he wrote no books, he sent several hundred letters in correspondence with the Royal Society in London to document his discoveries. The Royal Society published them, therein establishing his notoriety and accreditation. But it took another century for medical experimentation in artificial insemination to take off.

In 1786, Lazzaro Spallanzani, an Italian priest-physiologist, equipped with an updated version of van Leeuwenhoek's microscope, was the first to prove that fertilization required both a sperm and an ovum. In his book *Experiencias Para Servir a La Historia de La Generación De Animales y Plantas* (*Experiences to Serve to the History of the Generation of Animals and Plants*), Spallanzani documented his frog-mating experiments in which he made oilskin trousers for the frogs to keep semen from reaching the released eggs. He reported that the male frogs continued to grasp the females during mating, even while wearing the barrier pants, but could not fertilize the eggs because the oilskin trousers blocked delivery. By collecting the frog sperm from the condom-like trousers and exposing it to the female eggs, Spallanzani conducted the first-ever in vitro fertilization.

Spallanzani moved on to the artificial insemination of a cocker spaniel to breed a litter of three puppies and experimented with cooling, freezing, and preserving sperm. He gained worldwide recognition for these and other findings, and over the course of the next 150 years, scientists built upon his research in their use of artificial insemination to accelerate animal husbandry.

By the 1890s, artificial insemination science had become increasingly advanced and was practiced quite widely throughout Europe and Russia in the selective breeding of not only horses and dogs, but also of several domestic farm animals, including sheep, rabbits, fowl, and dairy cattle. Couldn't they be bred to be faster, stronger, cuter, friendlier, fatter, fluffier, or faster-growing? To be better swimmers or hunters? To produce more milk, more wool, or whatever?

Russian scientist Ilya Ivanovich Ivanov (1870–1932), was the first to document practical methods for animal artificial insemination. He perfected artificial insemination in horse breeding; his methods enabled one stallion to fertilize up to five hundred mares, as compared to twenty to thirty that could be naturally fertilized. Ivanov's process became the preferred practice used by equine breeders the world over and soon after became the standard practice used by domestic farm animal breeding in general.

Ivanov's work on crossbreeding and hybridization opened up innovative commercial applications such as the creation of new breeds that had the capacity to weather harsh Russian winters and that had increased resistance to sickness and disease. He applied his techniques to the preservation of various endangered species, such as the European bison, the wisent. His breeding of a horse and a zebra produced the only wild horse hybrid still in existence on the Russian plain.

Ivanov crossed the imaginary line drawn by both church and state, which would ultimately land him in exile. His transgression: bringing his experimentation to the human species. He had drawn the attention of the state as early as 1910 when he presented to The World's Congress of Zoologists the possibility of creating a human-ape hybrid through artificial insemination. Throughout the 1920s, Ivanov conducted experiments using human sperm to artificially inseminate three female chimpanzees. That effort failed. Reproductive science had not yet advanced to uncover the chromosomal limitations to hybrid life.

In 1929, Ivanov organized an effort to artificially inseminate human volunteers with orangutan semen, which ultimately led to his arrest. He was sentenced to five years in exile, starting in 1930. While in exile, Ivanov died of a stroke in 1932. His behavioral scientist friend and colleague, Ivan Pavlov, wrote his obituary and delivered his eulogy. History best remembers him for his nickname, "Red Frankenstein."

The more I learned about artificial insemination science, the more determined I became to learn about my own origin. I was starting to understand that, had it not been for strong-willed, intellectually passionate people willing to challenge conventional wisdom, scientific advances that either benefited or challenged social mores might not have occurred and my conception might have never been.

While I was on this journey of discovery, I shared my findings with my close circle, including my best friend, Mike. A one-time pre-med student whose lack of organic chemistry acumen flushed that dream down the proverbial drain, he was an engaged audience.

"Fortunately for me," I jested, "the human-ape hybrid experiment of Red Frankenstein failed."

He joked back, "After knowing you for so long, I'm not so sure. There's no telling what your paternal origins are."

Thanks, friend!

I later called him when I read about the opening episode of the 2011 sci-fi television series *Dark Matters: Twisted but True*, entitled "Ape-Man Army," which dramatized Ivanov's man-ape hybrid.

"You should have auditioned for the part," he guffawed.

Reproductive science initially hatched in the plant and animal kingdom. The next step in my research project was to understand how it had been applied to humans, and ultimately to *me* specifically. What was my origin? Was I a Frankenstein like product? To fully process that, I wanted, and I needed, to understand the whole background story of how I was hatched. I found that human reproductive experimentation, as well as the spread of some legends about it, moved concurrently with the experimentation of scientists working with flora and fauna.

Chapter 7

My conception wasn't a Frankenstein-like experiment.
I continued to repeat these words to myself again and again as I learned more about the biological science of artificial insemination. Instead, I thought of myself as the product of a science-perfected practice from a fertility clinic. My parents wanted children, and my dad was sterile. It was that simple, but the history of what led them to be able to have me was complex.

Before the end of the eighteenth century, the only way for a woman to conceive a child was *au naturel*. Those desperate to conceive might have enlisted a stand-in for an infertile husband—which was adulterous, dangerous, and socially unacceptable behavior. If discovered, the resulting child would be labeled a "bastard" and expelled from any patriarchal legacy. And the mother? Abandoned, at best.

Adultery and bastard stories are as old as the Old Testament itself. To really understand the background of my conception, and the secrecy behind it, I wanted to understand the cultural context and origin of the secrecy and the biases that contributed to it. I discovered that in Old Testament times, adultery by a married or betrothed woman was considered a crime against her husband or fiancé on the part of both participants and also considered an evil act, subject to the death penalty:

> If a man is found lying with the wife of another man, both of them shall die, the man who lay with the woman, and the woman. So you shall purge the evil from Israel. If there is a betrothed virgin, and a man meets her in the city and lies with her, then you shall bring them both out to the gate of that city, and you shall stone them to death with stones, the young woman because she did not cry for help though she was in the city, and the man because he violated his neighbor's wife. So you shall purge the evil from your midst.
>
> —Deuteronomy 22–24, New International Version

The history I had learned in school also stuck with me. The ancient Roman and Jewish worlds were renowned for their lax sexual morality, at least according to modern standards. Perhaps that was the reasoning behind the strict laws of the early Christian church: "Christians must not associate with those who are sexually immoral" (1 Corinthians 5:9, New International Version).

Galatians 5:16–21 (New International Version) is more descriptive: " . . . do not gratify the desires of the flesh." It lists the "works of the flesh . . . fornication, impurity, licentiousness, idolatry, sorcery, enmities, strife, jealousy, anger, quarrels, dissensions, factions, envy, drunkenness, carousing, and things like these."

Notice that sexual immorality is mentioned first, highlighting its importance.

Even so, in the New Testament, Jesus preached compassion and forgiveness for someone who truly regretted a sin, as is shown in the story of the adulterous woman who anointed Jesus's feet: "I tell you, her sins, which were many, have been forgiven" (Luke 7:36–50, New International Version).

Adultery is one thing. If it resulted in a child, that was another.

The word bastard is associated with "polluted" or "foreigner," denoting those who do not share the privileges of God's children. The Bible is rife with stories of how Satan used the bastard curse to frustrate God and keep Him from enjoying close fellowship with His people.

As time evolved, I wondered to myself what constituted being a bastard. I began to research the history of illegitimacy.

Anthropologists agree that while attitudes have differed from society to society, illegitimate children have nearly always carried the bastard stigma. Thanks to my parents' finagling, my twentieth-century American society had considered me "legitimate," though I was positive that donor-conception had no place in the earlier centuries of Western civilization. I studied some more.

My research led me through European Medieval Christian and legal history, and I landed squarely in what was to become known as the nation of Spain. Prior to the thirteenth century, legitimate marriage was not the primary factor in determining a society's perception of a child's birth. Rather, the social status of both parents, the "right parents," carried the information needed to determine whether a child would inherit land, property, or title. For instance, William the Conqueror, sometimes referred to as William the Bastard, was born to Robert, Duke of Normandy, and Herleva, who was not his wife. But the duke recognized William as his heir. William conquered Normandy and England and passed his titles and kingdom to his children.

Society in the Middle Ages measured its leaders based upon their pedigreed ancestry and the power associated with those ancestors. Only beginning in the second half of the twelfth century did birth outside of lawful marriage begin to render a child illegitimate, ineligible to inherit a noble or royal title.

In the 1160s, following the death of William de Sackville, his nephew, Richard of Anstey, filed a legal claim against Mabel de Francheville, William's only child, for the recovery of his uncle's lands on the grounds that Mabel was illegitimate. William de Sackville had married Albereda de Tresgoz, with both parties consenting to the marriage, before he married Mabel's mother Adelicia. Mabel contested Richard's claim in court, asserting that a daughter's inheritance trumped a nephew's claim of inheritance. A ruling in the name of Pope Alexander III declared Mabel, a child who had resulted from her father's second "illegal" marriage to Adelicia, illegitimate.

This appears to be the first time an individual's inheritance was barred because her parents had an illegal marriage. The Roman Catholic Church voiced an influential opinion, but the overriding victory belonged to clever Richard of Anstey, who exploited theological doctrine to establish legal precedence. After that time, more and more plaintiffs began to do the same.

This history of inheritance surely did not apply to me. There was nothing to inherit! And if any donor-conceived person attempted to lay claim to an inheritance from a known donor, legal precedent surely stood in the way of any such claim of an "unnatural" child.

In the eyes of Roman Catholic Church marriage law in medieval Christian Europe, any child conceived by a couple not legally married was a bastard child, with no legal standing in society. As Christian Europe expanded into its varying empires, its practices, norms, and laws spread throughout its colonies. Each colonial law regarding adultery and its views on illegitimacy took its cue from old English common law. Common law reflected decisions applied throughout a region. Once a judge had come to a decision on a particular case, a precedent had been established for future decisions on similar cases.

Initially, common law regarding inheritance and legitimacy concerned itself only with the affluent. The bastards of the affluent were generally taken care of by their birth mothers, with financial support provided by their fathers (if the women were lucky). In 1531, the first legal statute related to bastardy, known as Henry VIII's Old Poor Law, defined a bastard as a "Fatherless Poor Man's Child," and thus the responsibility of the community. Fatherless children of poorer families survived on handouts from relatives, food and shelter provided by monasteries, and financial donations from a charitable population. Also in 1531, Christ's Hospital was built in London with specific instructions to provide care for bastard children.

By 1574, with a large number of bastard children overloading the monasteries and municipalities, an update to the statute created a new legal precedent that required that the father of a bastard child be financially responsible for his child. Should he not pay, the mother had the right to have him arrested and even jailed. Society provided the support funds in such cases. That law also expanded the power of municipalities to raise taxes to care for the poor.

To discourage adultery and promiscuity, further refinements to The Poor Law (codifying the blatant sexism common at the time in which women were held accountable but the men were not) established that the mother of any bastard child would face corporal punishment or be placed in an English "House of Correction." Using this common law, colonial courts not only fined those found guilty of fornication or adultery, but also added the humiliation of public whippings.

Throughout the seventeenth and eighteenth centuries, the rising cost of funding the support of illegitimate children (who were increasingly stigmatized) and forced marriages to offset that cost alarmed Victorian England and further fueled a reformation of its Poor Laws, which were deemed by a reviewing commission to "encourage licentiousness and illegitimacy." In February 1834, the *London Times* editorialized that poor relief be reserved for the destitute and that the promiscuity of the mothers of illegitimate children be punished. (Demonstrating once again the sexist focus on mothers and the blind eye turned to the fathers of such children.) The Lord Chancellor in the House of Lords denounced "the lazy, worthless, and ignominious class who pursue their self-gratification at the expense of the earnings of the industrious part of the community."

Additional refinements to the Poor Laws in 1834 remedied what was considered decaying morality. New statutes absolved fathers from any responsibility for their bastards, thus socially and economically victimizing the mothers, who had sole responsibility for the support of their children to the age of sixteen. If the mothers were unable to support their children, they both would have to enter the workhouse in their parish. The social and economic ostracism of those women was meant to inspire virtue and deter the rise in illegitimate birth; instead, it furthered poverty, promoted infanticide, and enabled a murderous business called baby farming to flourish.

Baby farmers collected a fee for the care of an illegitimate child—the younger the better. This essentially freed the mother to start anew and seek a better life. More children meant more fees. The children given over to the baby farmers were either poisoned or starved to death. I pondered how modern society would have addressed this infanticide. Then I realized my answer: abortion.

It was not until 1889, with the help of what is today known as the National Society for the Prevention of Cruelty to Children, that reforms to the Poor Laws protected infant life and provided financial support for illegitimate children. Under the changes to the law, fathers bore equal responsibility for their illegitimate children until the age of sixteen and Poor Law boards could provide aid to the mother. When I learned that the National Society for the Prevention of Cruelty to Children was formed a full sixty-five years after the Royal Society for the Prevention of Cruelty to Animals, I could not help but note that civilization at the time held its animals in higher regard than its bastard children.

The "bastard curse," as etched in Judeo-Christian culture, was deeply ingrained in world attitudes by the time human artificial insemination came into practice. I was about to confirm my hunch that this classification applied to donor-conceived children too. I had enough baggage to carry. Now I had to add the bastard label to the equation?

When I talked about this with Mike one night after a dinner out, he came to the rescue with a cognac and a much-needed reality check. "You're not a bastard," he said, and then took a swig. "Sometimes a prick, yes! You're a CEO. But a bastard? No, I don't think so."

Mike always knew how to rein me in when my overthinking and anxiety went into overdrive.

Feeling more grounded, I was ready to dive into learning more about the actual evolution of the practices of human artificial insemination. I did not expect to find it had been an easy road of social acceptance for the pioneers. A venture capitalist phrase immediately resonated with me: "Pioneers often wind up with arrows in their backs."

Early artificial insemination by donor pioneers had collected their share of arrows soaked in venomous criticism from a skeptical society.

While science had not quite figured out how conception worked, the first recorded attempt at human artificial insemination was dubiously claimed in 1462 by Enrique IV of Castile, later dubbed Henry the Impotent. Born in 1425, he was the heir-apparent of John II of Castile. His mother was the daughter of King Ferdinand I of Aragon.

The Spanish and Portuguese aristocracy had waged war and engaged in power struggles and intrigue throughout his life. Fifteen-year-old Henry married Blanche of Navarre in 1440 to fulfill an arrangement made by his father in 1436 in peace negotiations between Castile and neighboring Navarre. The marriage dowry included Navarre territories and villas won by Castile during a war. King John II agreed to return the captured lands provided Navarre would gift them back to Castile as part of this dowry.

In 1453, after thirteen years of marriage, just prior to his succession to his father's throne, Henry sought a divorce, claiming that a marital curse which rendered him temporarily impotent (Blanche was his maternal cousin) had prevented consummation of their marriage. An investigative priest questioned local prostitutes, who testified that Henry was sexually capable. Blanche was confirmed a virgin after an official examination. The bishop annulled the marriage and sent Blanche home.

When he was crowned King Henry IV, succeeding his father in 1454, he was free to gain another beneficial alliance. Henry married Joan of Portugal, daughter of King Edward of Portugal and sister to Afonso V of Portugal, in 1455.

Political intriguers waged an unrelenting campaign to diminish Henry's power and influence throughout his reign. Rumors that he was impotent or a gay man grew as he spent his time away from his wife in favor of the Royal Balthazar in Madrid. When she unexpectedly gave birth to a daughter, Joanna La Beltraneja, rebel nobles claimed that the new princess and heir to the throne was actually the daughter of a neighboring duke. Queen Joan's subsequent adulterous affair, which produced two children, provided additional evidence used to discredit Joanna's legitimacy.

Royal chronicles of Henry IV's reign were written and rewritten several times under competing influences, but those loyal to Henry put forth a version of events that included the delivery of his life-giving seed in absentia, thus making Joanna a legitimate heir.

There it was. A claim of artificial insemination by husband. The first noncoital fertilization of a human being. Not preposterous! If the Arabs could inseminate their horses remotely a century earlier, so could a king artificially inseminate his wife.

I found the claim to be creative for the fifteenth century. Had the advances in science that were to come in later centuries been available at the time, and had the Roman Catholic Church been less oppressive in its views on sexuality, adultery, and legitimacy, Henry IV's account of his daughter's birth might have captured a more sympathetic audience, and history would have been substantially altered.

Of course, the quest for riches and power can corrupt human behavior—a concept that is as true in today's political and corporate worlds now as it was back then. Henry's younger, politically savvy, and ambitious half-sister Isabella had her own ideas to enable her succession. She influenced the League of Nobles with promises of their own increased riches and power under a national unification strategy and enabled the writing of another version of the Royal Chronicles to discredit Joanna's legitimacy as heir to Henry IV's throne. Henry ultimately divorced Joan after her lengthy and indiscreet affair with the bishop's nephew. After years of skirmishes between the rival factions, he finally agreed to name his half-sister Isabella as his successor.

After Henry died in 1474, Isabella became one of the most powerful and influential European monarchs. She united the nobles who had ruled what were previously disparate provinces, initiated powerful political and military alliances on behalf of Spain, entered into profitable trade agreements, and even funded the voyages of Christopher Columbus to discover valuable new trade routes in the New World. During her reign, the International Federation of Chess, the governing authority of the game, restructured the rules. While the king remained the most important, the queen (modeled after Queen Isabella), moved freely in any direction and became the most powerful piece on the board.

The origins of reproductive science appeared to be a chess game unto itself, requiring moves, sacrifices, strategy, and forethought. How had my life-giving sperm been delivered? Not in absentia by the man I knew as my father.

Henry IV's assertion of the first man-to-woman artificial insemination in 1462 had captured the imagination of a new generation of reproductive scientists. Their advanced discoveries gave me life with a caveat, a cloak-and-dagger secrecy attached to my conception. I repeated to myself: "No more secrets!"

A breakthrough in artificial insemination came at the end of the eighteenth century. For centuries, infertility was believed to be a woman's problem—her physical deficiency. Only a few medical authorities ever considered the possibility of male deficiency. Then, in 1790, Dr. John Hunter made reproductive science history.

Hunter, a Scottish surgeon and noted researcher deemed "the father of modern surgery," and official surgeon of England's King George III, contributed notably to advances in medicine. He helped to improve our understanding of human teeth, bone growth and remodeling, inflammation, gunshot wounds, venereal diseases, digestion, the functioning of the lacteals, child development, the separateness of maternal and fetal blood supplies, and the role of the lymphatic system. But he is most acclaimed for the first recorded human artificial insemination effort.

Dr. Hunter treated a young couple desperate to have a child. The husband suffered from penoscrotal hypospadias, a severe birth defect in boys in which the opening of the urethra is not located at the tip of the penis, but rather where the penis and scrotum meet.

Dr. Hunter equipped his fertile, but anatomically deficient, male patient with large syringes and instructed him to masturbate frequently, collect his semen in a warmed syringe, and inject it directly into his wife's vagina. While their bedroom intimacies were never documented, the couple essentially followed his instructions, persevered over two years, and finally conceived a child.

Although this successful conception occurred late in the eighteenth century, it was 1866 before the "father of modern gynecology," American surgeon J. Marion Sims, entered the artificial insemination scene with controversial methods that yielded groundbreaking discoveries.

Dr. Sims, I mused. There he was. He wasn't my mother's Dr. Sims because his dates were in the mid-1800s, not 1945. *Perhaps a relative,* I thought.

Chapter 8

Both before the Civil War and long after, Dr. J. Marion Sims was a controversial figure. I learned that he was professionally revered as a visionary innovator in the nineteenth century and rightly discredited as an appallingly unethical racist in the twenty-first century.

Over the course of his professional career, Dr. Sims chalked up numerous discoveries of aspects of female anatomy, pregnancy, childbirth, and surgical procedures. His initial acclaim had derived from developing a surgical procedure that repaired vesicovaginal fistulas—tears in the vaginal wall that often occurred during childbirth and caused incontinence.

Dr. Sims had no formal background in gynecology prior to beginning his practice. He worked as a plantation physician in Montgomery, Alabama, throughout the 1840s in a backyard hospital that he had opened. Although his experiments led to an increase in gynecological knowledge, the information came at a terribly high price paid by the victims of his experiments. He conducted experimental surgery on enslaved women who were forced by their enslavers to undergo his experimental procedures. Although anesthesia had recently become available, he did not give it to the enslaved women during the surgical experiments—his approach was akin to the way Nazi doctors withheld anesthesia from the Jewish prisoners on whom they experimented during the Holocaust. His methods were vociferously and justifiably discredited once they were fully uncovered in his own journals nearly 130 years after his death.

In 1855, armed with his newfound anatomical knowledge gained at the expense of the pain and suffering of a dozen unsedated Black women who had undergone multiple surgeries in his backyard hospital, Sims relocated to New York City. There, he opened the Women's Hospital, the first and only one of its kind dedicated to women's health; the hospital was sponsored by wealthy New York women. Sims was ridiculed and called a quack and a fraud by other physicians who claimed no need for that specialty existed. At the time, established physicians considered gynecology ("the practice of examining the female organs") to be repugnant. Medical schools often used dummies to train student doctors on how to deliver babies. New doctors generally encountered their first female patient only once they began their own practice.

Dr. Sims often performed surgeries in a surgical amphitheater so medical student and doctors could watch and learn his technique. The Women's Hospital went on to serve the poor and disenfranchised in New York City who could not otherwise afford treatment. During the mid-century wave of Irish immigration during the Potato Famine, indigent women found care at the Women's Hospital that they otherwise could not have afforded. A full one-third of the immigrants coming to the United States at that time were from Ireland.

I did not have any Irish ancestry—that I knew of at this point anyway. But I wondered whether some of his patients had been Susan's relatives or the relatives of my friends with working-class, Irish immigrant backgrounds.

While the hospital continued treating the poor and disenfranchised under his direction, Dr. Sims altered the thrust of his personal practice after being summoned to Europe in 1863 to examine Empress Eugenie, the wife of Napoleon III, to treat her fistula. With his worldwide reputation as a gynecological surgeon at its peak, he returned from Europe, raised his rates, and limited his private practice to the type of affluent women who had funded the Women's Hospital.

After twenty years of building up the Women's Hospital and earning a reputation for being difficult to please, cantankerous, and a braggart, Dr. Sims was dismissed by the hospital board in 1875 due to his insistence on treating uterine cancer patients. The prevailing attitude of the board and the majority of the hospital's physicians was that cancer was not specific to women. Plus, at the time, the larger medical community thought cancer was contagious. They were fearful of catching that generally terminal disease.

Adding to the conflict, several members of the hospital board attempted to restrict Dr. Sims's technique of teaching in an amphitheater in a way that "exposed the genitals of white women." Although his dismissal occurred after the Civil War and the abolishment of slavery, racism still pervaded White society. Sims would not compromise. The various conflicts all added up to his dismissal.

The medical community was aghast. The world-acclaimed father of modern gynecology let go by his hospital board? Outrageous! His dismissal was regarded as an example of political infighting due to disagreements in strategy that had nothing to do with his competence.

To recognize him and make a supportive statement to the board of the Women's Hospital, the American Medical Association named Sims its president from 1876 to 1877. He then set in motion the founding of his second New York City hospital, the New York Cancer Hospital (now known as Memorial Sloan Kettering Cancer Center). It was the first such American institution dedicated to treating cancer patients.

Its grand opening occurred shortly after his untimely death in 1883 by cardiac arrest.

Dr. Sims's notoriety reappeared like a bad dream twenty years after I had completed my initial research into his life. A statue to honor him and his discoveries had been erected in 1890 at East 103rd Street and Central Park in New York City, across the street from the New York Academy of Medicine. The statue was the target of innumerable protests when people became outraged over his cruel methods, which became public knowledge in 2016. His statue was finally banished from its location in 2018. The initial plan for the statue was to relocate it to a more obscure location—his gravesite at Brooklyn's prestigious Green-Wood Cemetery. But through 2020 it has remained in a cemetery warehouse.

While I found Dr. Sims's professional biography quite a fascinating read, his reproductive science experiments captured my attention most because they gave some substance to my own conception. In New York City, Dr. Sims conducted fifty-five artificial insemination experiments on six of his affluent patients and reported the results in various medical journals in 1866.

Sims's version of artificial insemination included a few faulty suppositions and practices that hindered his success. First, he always assumed that infertility was caused by some anatomical problem in the woman, oftentimes a cervical or tubular obstruction or misalignment that could be corrected by surgery. After all, he was an accomplished surgeon who had had several successful experiences with this type of corrective surgery.

Second, he always used the sperm of the woman's spouse, never that of a third party. Again, the general assumption and social attitude was that men were always fertile. Infertility was a woman's issue. And as history has shown, society would likely have had issues with an "adulterous" use of third-party sperm to impregnate another man's wife.

Third, he believed that ovulation occurred during or very close to menstruation. The scientific identification of the exact timing of ovulation had yet to evolve. Since antiquity, it was thought that breast tenderness and ovulation pains pinpointed its timing. Dr. Sims subscribed to the conventional wisdom that fertility coincided with the beginning of the menstrual cycle.

Even if the women were highly fertile, and the spouses' sperm were potent, it could be rationalized that the women were not likely ovulating and hence not impregnable when he inserted the sperm.

Sims documented all his findings in voluminous published scientific writings detailing his medical experiments. Along with his self-aggrandizing 471-page autobiography, those writings have been the main sources of knowledge about him, his career, and his unethical methods. The entire practice of gynecology and fertility treatment was built on his discoveries. Modern society has compared Dr. Sims to the Nazi physician Josef Mengele, who is known as "the Angel of Death" for his deadly experiments on Jewish prisoners held captive in Auschwitz.

During a 1996 annual physical exam with my Boston-based primary care physician, with whom I enjoyed an amiable rapport, I highlighted my donor-conception discovery, my mother's clues, and some details of my early research.

The doctor, whom I simply called Harold, recognized Dr. Sims right away. "Oh, yes, the father of modern gynecology. That was too long ago. He's not your guy."

Before leaving his office, he handed me a slip of paper. Written in his atrocious handwriting was a name, Dr. Greene, and his phone number. "Here's a guy I know with the American Medical Association," he told me. "Give him a call and let him know I gave you his number. Perhaps he can help you."

I was excited. I called Dr. Greene that afternoon. After not getting a response, I called again a week later. My voicemail messages were never returned. Disappointing! For a moment, I thought I was closing in on the knowledge I sought. I was gathering enough information to write a book, but what I really wanted was a tangible clue to my genetic origin.

I discovered that Sims had a son who had also practiced medicine—another Dr. Sims. But the younger Dr. Sims died four decades before my parents were even married. He was also based in New York City, not Boston. Who was my parents' Dr. Sims? I continued to hit brick walls with my mother's clues.

My mother, even during her pre-stroke days, was almost comical in her penchant for misremembering names. My family could all agree on this. She would confuse someone's name with another person that she met at the same place or around the same time and would construct a hybrid name that combined the two. She often laughed at herself.

For instance, while I was in high school, I began an enduring friendship—little brother, big sister—with a girl named Suzanne. Blond, blue-eyed, and attractive, she was two years older than I and was also an only child. While feminine, she had a unique personality: She sailed a boat competitively, drove her father's double-clutch pickup truck, and rode a Vespa.

As a sophomore, I was popular with the junior and senior boys who would often ask me for an introduction to Suzanne. But I had her back. For a time, we were inseparable. A decade later, I married a woman named Susan. I always corrected my mother when she referred to my wife as "Suzanne." Finally, to accommodate the correction, since she couldn't remember who was called what, she referred to both Suzanne and Susan as "Sooze." Now, in trying to identify the source of the sperm donor, we were dealing with her post-stroke recollection of someone's name.

I questioned my mother further. "Is it possible, Mom, that your fertility specialist was building on the research done by Dr. Sims? That he was not actually called Sims himself, but had mentioned that name to you somewhere along the way?"

She wavered a bit then repeated somewhat tentatively, albeit convincingly, that her fertility doctor's name was Sims.

I turned back to my research to try to uncover another medically inclined descendant, but I found no such relative alive at the time of my conception in 1945. (Although I did learn that John Wyeth, the acclaimed poet and artist, was the grandson of Dr. Sims. John's father, also a southern doctor, had moved to New York City after serving as a Confederate surgeon during the Civil War. He met and married one of Dr. Sims's daughters. Because of the connection with my research, I have become particularly interested in Wyeth's WWI sonnets and his paintings.)

Dr. Sims's articles inspired the work of an 1800s reproductive entrepreneur, Dr. Edward B. Foote. An author and early proponent of birth control, Foote published a home medical advice journal called *Medical Common Sense*, which spoke frankly about sexual health. This journal not only described artificial insemination, but advertised an "impregnating syringe" by mail order for home use. While history is quiet about the success of his mail order business, it does pique the imagination, doesn't it? (Incidentally, do-it-yourself artificial insemination kits are available today on Amazon.com.)

After being convicted under the Comstock Act of 1873 (an "Act of the Suppression and Trade in, and the Circulation of, Obscene Literature and Articles of Immoral Use") for publishing birth control information, Dr. Foote founded the Free Speech League.

Neither Dr. Sims's nor Dr. Foote's experimentation with artificial insemination created social or theological outrage. The whole topic of infertility was, rather, confined to the scientific and medical community, which greeted it with an accepting curiosity until a credentialed professor at Harvard Medical School penned his negative views on the topic.

Dr. Edward Clarke's bestselling 1873 book, *Sex in Education; or, A Fair Chance for Girls*, cast doubt on the physiological origins of infertility, and his suspicious attitude spilled over to artificial insemination. He deduced that it was primarily the education of women, not emotional or physiological circumstances, that negatively impacted their fertility. Their wombs shrank as their brains filled with knowledge. He offered proof: Educated women who sought to exercise some control over their bodies had fewer children.

By the end of the nineteenth century, Clarke's influential views colored a developing public, legal, and religious discourse. The women's movement of the twentieth century seriously challenged the suppression of women and generated improvements to a sweeping array of social and human rights issues including voting, education, leadership, birth control, abortion, and fertility practices.

Secrecy and suppression had been integral to the culture of donor-conceived artificial insemination from the very beginning. One disciple of Dr. Sims concealed his work because, at that time, society would have considered it akin to the crime of rape perpetrated on a defenseless woman by a gang of villainous scoundrels.

Chapter 9

Just a year after Dr. Sims's death, in 1884, at Philadelphia's Jefferson Medical College (Dr. Sims's alma mater), with questionable or even criminal ethics, Dr. William Pancoast pioneered the first third-party donor-insemination procedure.

As a medical professor, Dr. Pancoast had taught Dr. Sims's gynecological surgical techniques and anatomical discoveries to his students. He had followed Dr. Sims's writings and was well aware of his predecessor's unsuccessful 1866 experiments in artificial insemination. Dr. Pancoast was also aware of scientific data on ovulation that had come to light two decades later. The position of the cervix was thought to be a tell-tale indicator of ovulation; it was to be lower in the vaginal canal, firm, and with a small opening during unfertile times and higher in the vaginal canal, softer, and more open during ovulation.

An older merchant with a much younger, long-barren Quaker wife enlisted Dr. Pancoast's help to conceive. Armed with more accurate scientific information about ovulation, Pancoast discerned that the husband's sperm was too old and weak to do its intended job. The doctor-professor asked to examine the wife one final time before he unloaded the bad news.

Dr. Pancoast then proceeded to conduct a completely unethical experiment on the young woman without explaining his intentions to her beforehand or acquiring her consent to the procedure. The "exam" took place as a classroom experiment with six medical students in attendance. The professor chloroformed the young wife, and the class voted on and selected the most handsome among them to masturbate and ejaculate into a syringe for Dr. Pancoast to inject into the unconscious woman. (If this conception scenario had played out a century later, I imagine both the professor and would-be doctor may very well have been [and should have been] convicted and incarcerated in a highly publicized case of rape-by-doctor or aiding, abetting, and covering up the crime.)

Success on the first try! Once the woman was obviously pregnant, Dr. Pancoast confessed to her husband the means of his wife's pregnancy. The doctor asked his handsome medical student to stand by in case the husband reacted violently. Reportedly thrilled with the result and spared the embarrassment of being deemed sterile, the husband agreed to keep the entire procedure confidential. The husband, steeped in the paternalistic and sexist attitudes of the time, insisted that the information be withheld from his wife and child—and thus he contributed to the cover up of the procedure performed on the young woman without her consent. The experiment resulted in the birth of a healthy baby boy.

Prior to Dr. Pancoast's experiment, artificial insemination had been confined in practice to the use of a husband's sperm, just as Dr. Sims had documented, and there are very few accounts from other doctors. After Dr. Pancoast's violation of the young woman medical knowledge about male fertility and sperm potency grew. A new experimental approach collected a husband's weak sperm in a laboratory. The sperm was examined microscopically to group the "strong swimmers" in an attempt to boost potency. The new, rather loose, term "test tube baby" caught on, describing a scenario in which a husband's semen had been inserted via an instrument through the wife's cervix to overcome whatever anatomical issue the couple was experiencing, hypospadias or a tilted and somewhat obstructed cervix among them. An uncredited medical student termed the procedure "a three-inch boost for a six-inch journey."

In 1909, after the death of Dr. Pancoast, the handsome sperm-donating former medical student, Minnesota physician Dr. Addison Davis Hard (you cannot invent these names), let the artificial insemination by donor cat out of the proverbial bag. He documented the entire scenario in a letter to a prominent journal, *Medical World*. Just prior to the publication of his letter, Dr. Hard presented himself as the biological father of the now-adult baby boy and unveiled the tale of the young man's conception. To prevent any embarrassment or unwelcomed publicity, Dr. Hard never identified the couple or the young man, nor did he disclose the details of their conversation. By the time of the doctor's revelation, the merchant father was deceased. It is unknown whether the young man's birth mother was still alive.

How might both the young man and his mother have reacted? Would they have believed such a story? Would they both have felt that they were victims who had been violated in some sort of scam? Were they angry?

I suspect that they dealt with an array of emotions that were made even more complex by the shroud of secrecy. My own mother perpetuated the lie about my conception until I confronted her with the story I had made up to camouflage a flaw that I believed I carried. I imagined a stranger knocking on my door and telling such a tall tale. How might I have reacted at age twenty-five—or the age of forty-nine, when my donor-conception was disclosed to me? I suspect I would have reacted in much the same way that I actually did. Because of all the emotional baggage I carried related to my father's mental health issues, I was relieved. But I was left with genealogical bewilderment.

Once Dr. Pancoast's secret was revealed, the young man at least knew his genetic origins. I felt a kinship with him. We were both conceived via a secret procedure. We both likely experienced a full range of emotions, from gratitude to anger to betrayal, as we absorbed the reality of the situation. And I felt envy; he had knowledge about his genetic identity that I might never have. Had he and his biological father maintained a relationship thereafter? What about any half-siblings?

But the prospects of donor and half-sibling relationships were not driving me. Rather, my ongoing research was motivated by the desire to know my heritage and my medical history.

Upon Dr. Hard's revelation about Dr. Pancoast's secret, groundbreaking procedure, the Roman Catholic Church quickly and forcefully condemned donor-insemination specifically, and artificial insemination in general. Unnatural, noncoital conception was against God's law and sinful; ejaculation outside the vagina was labeled masturbation.

I recalled again some of my old history lessons.

It was 1909 when Dr. Hard disclosed the first artificial insemination by donor. Although the early twentieth-century Roman Catholic Church still exerted a heavy influence on society, various innovations had disrupted its power. The Industrial Revolution was disrupting the old-world economic order, and the melting pot of immigration had inflamed social and ethnic elitism. The 1909 Roman Catholic Church was still playing by medieval rules that had impacted Henry IV, but the rules of society had been changing drastically. As artificial insemination gained scientific and medical traction, religion, law, and society played catch-up to its reality—but they grew and changed at a snail's pace.

Dr. Hard's *Medical World* article stimulated a groundswell of disclosures about artificial insemination by donor. At the outset, a handful of claims emerged in medical journals and scientific writings from doctors who had been practicing artificial insemination by donor. Claims continued for over two decades, not only in the United States, but also in Europe, with more in-depth disclosures from a few practitioners in France and Germany.

Those claims merged with the timing of a global surge in the modern eugenics movement that grew to its height in the early twentieth century. "Eugenics" has a Greek origin: "eu" for good or well, and "genos" for offspring. The ancient Greeks and Romans practiced infanticide to selectively create a population with more desirable citizens. So, too, had the elite class in Plato's *Republic* advocated for "judicious mating" in couples with "the same natural capacities." The modern practice of eugenics sought to influence birth rates to selectively create children with the "correct" hereditary traits that would, in the view of proponents, benefit society.

Eugenicists promoted policies that encouraged preferred births and discouraged reproduction of those with "bad" genes via sterilization, either incentivized and voluntary or involuntary and compulsory (such as the 1913 Mental Deficiencies Act of the UK), or even genocide. It also promoted what was termed "miscegenation" marriage laws to prevent interracial unions (i.e., the Racial Integrities Act of Virginia). Eugenics practitioners sided with Social Darwinists, who advocated "survival of the fittest." From their perspective, for example, if there were no welfare policies, the high mortality rate of the reproducing and undesirable poor would positively impact society.

The eugenics movement went into overdrive under the influence of biologist and social scientist Charles B. Davenport (1866–1944). Considered the "father of modern eugenics," his first book, *Eugenics: The Science of Human Improvement by Better Breeding*, became the first major bioengineering study that linked plant development to human reproduction.

Davenport's advocacy spread the eugenics movement worldwide. With funding from the Carnegie Institute, he founded the Station for Experimental Evolution in 1904 (later renamed the Carnegie Department of Genetics) and the American Breeders' Association (ABA), the first eugenic body in the United States, in 1906. The ABA was formed specifically to "investigate and report on heredity in the human race and to emphasize the value of superior blood and the menace to society of inferior blood."[1] He attracted the initial financial support to enable the founding of the Eugenics Records Office in 1910 and became the first president of the International Federation of Eugenics Organizations in 1925.

Davenport's views, which historians consider outrageously racist, white-supremacist, and anti-immigration, influenced social and immigration policies throughout the world; those policies were aimed at attracting the "right" new population. Those views, for instance, later became embedded in US law in the form of the Immigration Law of 1924, which restricted immigration from southern and eastern Europe. Davenport's views had three major impacts.

First, immigrants from eastern and southern Europe who came to the United States throughout the late nineteenth and early twentieth centuries were seen as diluting a pure population. During that time, for eugenics disciples, "Mediterranean" was a negative term; "Nordic" was positive. I was certain that my dad had faced his share of Italian immigrant bias from this "pure population." My birth was the catalyst that led him to legally alter his four-syllable family name that was laden with vowels.

The date on my Change of Name Certificate is just five weeks after I was born. I was his son. I would carry his name for the rest of my life.

---

1   W. D. Stansfield, "The Bell Family Legacies," *Journal of Heredity*, Volume 96, Issue 1 (January/February 2005).

I recall him commenting when I was a child that he might not have done so if his wife had given birth to a girl. He carried his generation's expectation that a woman would marry and take her husband's name. I fantasized that if I had learned about my donor-conception as I began my adult life, I might have changed my name yet again. Guidaboni translated to "good guide." "Guide" (and a good one) would have suited me. Perhaps I would have tweaked the spelling to denote the common English or French heritage on both sides of my tree. *Easier still,* I thought, *would have been adding an accent to the end—same spelling, more French.* It would have been *my* name, *my* ethnicity, and not a half-name belonging to someone else.

Second, eugenicists perceived a "race suicide" caused by the declining birth rate in the Christian world that added to the world's population imbalance. Statistically, population growth in the less economically prosperous, less educated, and non-Caucasian countries in Asia, Africa, South America, and the Middle East far outstripped the lower growth rates in North America and western Europe.

Third, corrective actions by the "deserving few" were needed to better balance the population, through either reproduction or immigration. The "deserving few" were intelligent, educated (or at the very least literate), and potentially powerful. And Caucasian.

No shortage of high-profile world thought leaders considered themselves Davenport followers. Among them were some that surprised and alarmed me, including George Bernard Shaw, Winston Churchill, H. G. Wells, Alexander Graham Bell, W. E. B. Du Bois, and Theodore Roosevelt, as well as the then-presidents of Harvard and Stanford Universities and Bowdoin and Wharton Colleges.

Proponents of eugenics and its discriminatory features had generated public discourse that impacted society's laws and policies. Even the Roman Catholic Church promptly weighed in to condemn eugenics as elitist, repressive, and racist. I became all the more alarmed when I discovered that eugenics had infiltrated attitudes toward and early practices of artificial insemination.

My research uncovered one notable 1917 Oklahoma artificial insemination practitioner who had integrated Davenport's philosophies into his practice of artificial insemination. Dr. Frank Davis was the former superintendent of the Oklahoma State Hospital for the Feeble-Minded. There, he treated, comforted, confined, and sometimes cured his "inferior" patients by sterilizing them.

Davis had caused modest heart palpitations in the medical community in 1917 after publishing his book *Impotency, Sterility, and Artificial Impregnation*. In that book, he advocated for the sterilization of "inferior" people and for reproduction by "superior persons" via *au naturel* insemination by husband or artificial insemination by donor methods. He had the ardent support of a powerful Oklahoma eugenics advocate, William Henry Davis "Alfalfa Bill" Murray. Murray, while Speaker of the Oklahoma Assembly, a US congressman, and finally Oklahoma's governor, was instrumental in the passage of legislation regarding sterilization.

Dr. Davis had documented his artificial insemination treatments of several patients, recorded his successes, and detailed how he had taught couples to practice artificial insemination by donor at home using "borrowed semen from a superior individual."[2] In this way, he was much like Dr. Foote, and his procedures were similar to the protocol used by the Russian scientist Ivanov for his artificial insemination practices in the barnyard.

2   Frank P. Davis, *Impotency, Sterility, and Artificial Impregnation*
      (London: Henry Kimpton, 1917).

Chapter 10

Was my conception bioengineered by a Frankenstein-like doctor to be one of the chosen? My imagination raged. Was my parents' fertility doctor a eugenics practitioner, or worse, a Klansman?

Mike continued to poke fun at my anxieties. "Better that than being a hybrid from an ape."

This whole premise of being bred like an Arabian stallion gnawed at me. For the first time since learning that I was donor-conceived, I felt like a commodity. I tried to make light of it, but the very possibility both amused and frightened me. Popeye's words, "I yam what I yam," echoed inside me as I contemplated moving my self-perception from "flawed" to "perfectly genetically engineered."

Until the 1920s, scientific and medical writings had a narrow audience. Everyday people were not recreationally reading medical journals. But the contents of those journals were seeping into public discourse. For instance, *Scientific American*, once a four-page weekly newspaper that highlighted newly granted patents, altered its business model to become a monthly magazine that not only covered the impact of recent scientific discoveries but also explored the social issues attached to them.

During the decade of the Roaring Twenties, society began to change. It was the age of Prohibition and women's suffrage. Organized crime exerted a strong influence on society, and the economy and stock market were robust. The period ushered in indiscreet challenges to some of society's stringent behavioral norms on several fronts—the role of women, birth control, and fertility research among them.

Dr. Robert L. Dickinson, another unapologetic eugenics advocate, put that agenda front and center and broke that discourse open. While Chief of Obstetrics and Gynecology at the Brooklyn Hospital, he gained stature within the medical community for innovating a number of practice standards, among them tying off the umbilical cord before severing it. In 1920, as the newly elected president of the American Gynecological Society, he recognized the emerging practice of fertility as a new specialty. His controversial approach to treating fertility issues created quite a brouhaha on several levels.

Trained in classical art prior to his medical studies, Dickinson applied his talent to the creation of detailed illustrations of sex, fertility, and reproduction in his many medical journal publications and textbooks (for instance, *The American Text-Book of Obstetrics for Practitioners and Students*, 1903, or "Toleration of the Corset," in the *American Journal of Obstetrics and Diseases of Women and Children*, June 1911). He broke new scientific ground in his painstakingly detailed sketches in which he recorded his patients' sexual histories and genitalia. His resulting power struggle with the Comstock Act for nearly a quarter century ended only after the law was amended in 1936.

In Dickinson's first presentation to the American Gynecological Society, in October 1920, he urged the membership to study and pool their experiences with what he had termed "artificial impregnation." His audience of fellow physicians was shocked but remained silent. At least in that moment.

In December 1920, Dickinson disclosed his decades of experience with artificial insemination by donor, which included diagrams in his draft of a how-to manual.

Dr. Dickinson also documented a pregnancy achieved using not a husband's sperm, but rather a donor's sperm, in 1890. While scant on detail, his account was seemingly the second documentation of artificial insemination by donor. And Dr. Dickinson's procedure was completed under far superior ethical circumstances than Dr. Pancoast's drugging and violation of his patient at Jefferson Medical College in Philadelphia six years earlier.

Dr. Dickinson very clearly stretched the boundaries of accepted medical ethics of his day. Skepticism over the social, legal, political, and religious ramifications of donor-insemination ran deep. The prevailing attitude shaping discourse was that artificial insemination by husband could be discussed, but artificial insemination by donor was far too controversial. The American Gynecological Society's board severely admonished Dr. Dickinson and firmly instructed him to take sexual and reproductive topics off the agenda. Those discussions belonged behind closed doors for the few who had dared to express an interest.

History repeatedly teaches us that science continues to advance in spite of restraints that society attempts to impose on it. So, too, were advances in artificial insemination by donor made. Dickinson needed another forum since he had been silenced by the American Gynecological Society. He founded the Committee on Maternal Health in 1923 (later renamed the National Committee on Maternal Health) intending to sponsor medical investigation on sexual issues that the medical community at large avoided: contraception, abortion, and infertility among them. With the committee as his platform, Dickinson launched his first artificial insemination survey.

The committee's 1924 world medical survey reported 123 donor-conceived children, as disclosed by practicing reproductive physicians scattered throughout the United States and western Europe. Dickinson was so enthused by these results—which indicated that artificial insemination by donor was gaining traction around the globe—that he discontinued his medical practice. He turned his attention toward raising fertility, birth control, and artificial insemination as legitimate topics for the medical community overall to discuss and embrace, citing the growing constituency of those specialty physicians around the globe. He earned worldwide recognition as a leading "sexologist," and provided inspiration to the sex researchers Masters and Johnson three decades later.

In 1927, in "Dear Abby" fashion, *The Journal of the American Medical Association* (*JAMA*) began providing answers to anonymous questions from the handful of early fertility specialists regarding the best practices for achieving artificial insemination. Whether by husband or by donor, artificial insemination suffered a large failure rate; it often took dozens of attempts. Yet the number of donor-conceived children increased to 185 cases in a 1928 follow-up survey launched by Dickinson's committee. Fertility as a medical practice had gained prominence; it was still small in absolute numbers but had reported a 50 percent increase in the number of children conceived via artificial insemination over the previous five years.

The medical community had made steady scientific advances in understanding how a woman's glandular and hormonal makeup impacted pregnancy success rates. First, in 1906, renowned researcher and surgeon Dr. John Morris, after decades of experimentation on rabbits, reported in the *New York Medical Record* the birth of a living child after an ovarian graft. Another research breakthrough (with mice) by Dr. Edgar Allen and Dr. Edward Doisy isolated estrogen, the primary ovarian hormone. Their discovery further advanced the study of hormones and the endocrine system.

By 1929, gynecological researcher Dr. George Washington Corner recorded the extraction and impact of progesterone, which prepares the endometrium for pregnancy after ovulation and triggers the lining to thicken to accept a fertilized egg. It also prohibits the muscle contractions in the uterus that would cause the body to reject an egg. While the body is producing high levels of progesterone, the body will not ovulate.

My mother recalled that my parents' fertility specialist warned that they might require several procedures to successfully conceive. There was one chance in six of success. By the 1930s, however, the timing and signs of ovulation had been discovered. She related to me the precise process she followed before making her insemination appointment: "The doctor told me to take my temperature and look for a modest rise and a slight discharge of vaginal mucous when I was mid-cycle. He confirmed my ovulation by checking the position of my cervix" (higher, softer, and more open).

Perhaps Dr. Sims would have been far more successful in his 1866 attempts at artificial insemination, I observed, had he understood more about the timing and signs of fertility. In 1945, my mother's artificial insemination by donor succeeded on its first attempt. Was it in late February or the first half of March? She did not remember exactly, but it was a cold New England winter's day.

Throughout the 1930s and into the 1940s, in stealth, medical fertility practices emerged as a unique specialty. They filled a sizable, previously unmet need for the 10 to 15 percent of couples who had difficulty conceiving.

I focused my research on artificial insemination by donor prior to the establishment of the first sperm banks. As a 1940s baby, I was among the first unique group of babies conceived by donor-insemination prior to the sperm bank/frozen sperm age. The successful freezing and thawing of sperm was a 1953 scientific phenomenon, which enabled the 1970s evolution of sperm banks.

I wanted to understand how they had done it. I learned there were two simple choices.

Donors generally opted out of the Ivanov practice reserved for uncooperative cattle—either getting a hole poked in their scrotum or their prostate pressed to produce a drop. The first reported artificial insemination by way of a testicular puncture on an impotent husband was attributed to one German doctor, Hermann Rohleder, in 1904. He wrote a tell-all book, *Test Tube Babies: A History of the Artificial Impregnation of Human Beings*, in 1921. Since it was written in his native German and specifically targeted the medical community, it was not widely read.

"Donation" was an anonymous live-fire exercise, likely "by hand." It took place either in or adjacent to the doctor's office to enable a swift delivery to the intended woman. At best, human sperm has a two-hour period of potency in the open air.

At the time of my conception, early sperm donors were primarily sourced from a pool of university medical schools. Medical school students could earn an extra $25 to $50 per deposit for masturbating into a vial. That equaled about $250 to $500 in 2000 buying power when I looked it up, and $400 to $800 in 2020 buying power according to an inflationary index. Meaningful money.

I queried the older generation in my family and friendship circle about 1945 prices: my mother; my Uncle Bill; my mother-in-law, Kitty; Mike's father, Andy; and my high school mentor, Owen.

"How far did a dollar go in 1945?" I inquired.

They noted that a top-tier hotel cost $3 per night in the waning days of WWII.

As a former psychology major, I made some mental connections. Pavlov and B. F. Skinner researched and taught the impact of rewards in behavioral and motivational psychology. The medical school students were not salivating in anticipation of food, but they were donating their sperm for genuine economic incentive. A thought crossed my mind:

Had I known about the option of donating sperm while I was a poor college student, perhaps I would have donated myself.

Those early donors were superficially screened by the doctor or a member of his staff to physically resemble the recipient's husband: eye color, hair color, skin, blood type, general features, and overall health. Doctors of the day preferred donors who were married or engaged to be married, thus lessening the risk of a donor with venereal disease. Anonymity and secrecy prevailed to protect the privacy of donors and recipients and avoid legal complications.

To better ensure anonymity and the practice of secrecy, less than one-third of those doctors kept records on either the donors or the children resulting from the use of artificial insemination by donor. Parents were instructed to tell no one. Not even the resulting child. Ever! Anonymity meant anonymity; the attending physician would never break his code of propriety.

Fertility practitioners instilled that code of secrecy into their patients. By keeping the procedure a secret, the sterile man would never suffer embarrassment, the woman would never be accused of adultery, the doctor would never be blacklisted as an accomplice to that adultery, the child would never be considered a bastard, and would-be blackmailers would never have access to this incriminating information.

I had found all of this to be intellectually fascinating. I was an enormously motivated, erudite researcher, but I was still not getting what I wanted. "England, Russia, Philadelphia, Holland, Italy! Where is Harvard?"

My mother had said it was "Dr. Sims from Harvard, with a Boston office on 10 Beacon Street." I had not uncovered a single historical fact related to the science of artificial insemination that connected to anyone at Harvard at this point. I had to adapt my search, so I broadened my hunt to include fertility in general, with a circle very specifically around Harvard University practitioners. One name surfaced.

In 1939, Dr. Gregory Pincus, a Harvard biologist, documented the world's first in vitro fertilization implanted into a surrogate womb. He extracted the egg of one female rabbit and fertilized it in a laboratory using the sperm of a male rabbit. He then implanted the resulting embryo into the uterus of a different female rabbit. Presto! A litter! I also learned that Harvard was quite displeased with Dr. Pincus for conducting this research and dismissed him.

Scientifically an achievement to be sure, with a negative reward, but the dots just did not connect. I needed some positive reinforcement to reward myself for my time spent and the inconvenience required to discover my missing genetic information.

I fought the discouragement caused by my lack of discovery, or rather, by knowing that I might never find what I was seeking. Even if I found the missing Dr. Sims, uncovering fifty-plus-year-old records that seemingly did not exist in the first place seemed unlikely.

My conclusion thus far was that the donor of the sperm used in my conception was a Harvard Medical School student who had graduated between 1945 and 1948. On an early fall afternoon in 1998, I again crossed the Charles River and visited Harvard Medical School. With the help of a listing of Harvard Medical School alumni and the library's Xerox copier, I inserted into my briefcase a bright yellow folder that contained an alphabetized list of Harvard Medical School graduates from each of those four years.

In that folder was the name of the likely source of my paternal seed. I was convinced of it.

Chapter 11

O ver the next few years, I traveled back and forth to Harvard
Medical School's library or visited the Boston Public Library so
frequently that I was on a first-name basis with several of their staff
members. They inquired, "Are you a professor researching for a book?"

Not quite.

This research project had become very personal, and I shared my discoveries freely with my inner circle. We enjoyed intellectualizing about
them and played "what if" games about my conception and gene pool.

Mike continued to rib me with his Red Frankenstein hybrid
theme. "Perhaps this is how you evolved to become the Neanderthal
that you are."

My approach to the complex has always been to compartmentalize,
then integrate. I thought, *Let's test some of these historical discoveries on my
mother and see what she has to say.*

It was not altogether surprising to learn that what she had to say
was not much.

The artificial insemination by donor history was much more meaningful to me than to her, which I learned when I inquired "Mom, did you know . . ." and then recounted the history I had learned. Her answers were telling.

"No, I didn't know," was her flat reply.

"Isn't that interesting?" I probed.

"That happened before I was born," she said, dismissing the topic.

After five years of off and on research, the mysterious Dr. Sims at a fertility clinic on 10 Beacon Street in Boston was nowhere to be found. My paternal gene pool remained unknown. None of that appeared to interest her. She had freed herself from the weight of her long-kept secret. Our relationship remained whole. She appeared quite satisfied.

But by 2003, my mother was struggling—her health had deteriorated further. One corrected leaky heart valve led to a second leak in another valve. But this time she was too old and frail to consider corrective surgery. Her doctor advised, "Live your life."

Aside from not really showing an interest in absorbing all the information I had been collecting, my mother mirrored society's attitudes toward artificial insemination. The very same attitudes that had driven my parents' secrecy.

I understood that secrecy intellectually. But it changed from an observation to an inflammation that burned a larger empty hole within me. With genealogical bewilderment raging within me, I confessed to Susan that I feared I would live the rest of my life unfulfilled in my quest to learn about my biological origins. I felt that I had been living a lie. And I couldn't understand how society could make it acceptable to raise a child based upon such a fabrication, a lie, about his genetic origin.

I was soon to discover why.

Next on my research checklist was the social context of the times leading to my conception. That context also had ramifications on the practice of artificial insemination and the practice of family law.

Doctor Hard's 1909 disclosure to *Medical World* instigated a debate over the moral, legal, and social implications of artificial insemination by donor within both the medical community and the popular press in the United States. The press had outlandishly labeled a fertility specialist's active role in the insemination and conception procedure as "adultery by doctor." In 1920, after Dr. Dickinson disclosed his 1890 artificial insemination by donor procedure, the medical community at large clammed up when solicited for comments by the press and deliberately did not discuss artificial insemination, especially donor-insemination. It was far too controversial. Soon, the court of law joined the court of public opinion.

In 1921, a court in Canada set a legal precedent in the case of *Oxford v. Oxford*. A woman sued her husband for support. She had conceived a child via donor-insemination during their separation. The husband's defense termed the wife "an adulteress." Justice Orde, who presided over the case, deemed that donor-insemination without the husband's consent was even more than sinful—the justice ruled that it was as adulterous as conceiving a child *au naturel* and grounds for divorce.

A British court followed suit in 1924 and advanced the precedent of adulterous donor-insemination. In *Russell v. Russell*, Lord Dunedin granted the husband a divorce on the grounds of adultery, even though he had consented to his wife's artificial insemination by donor. The court also deemed the resulting child illegitimate.

It occurred to me that the power of legal precedent added a feverish urgency to the code of secrecy that my parents and their fertility specialist had practiced. My mother was satisfied, content that her child had never faced the social and legal hurdles of being declared a bastard and that she had never been declared an adulteress.

The world's religious, political, and social leaders all piled on to discredit artificial insemination as unnatural and unwanted. Would artificial insemination by donor encourage eugenic government policies?

What about regulation to prevent two biological siblings from the same sperm donor from marrying each other?

As the debate raged, the Archbishop of Canterbury declared that a donor should be held criminally liable for his adulterous participation. Like an avalanche, on a state-by-state and country-by-country basis, pre-WWII courts mostly ruled that a donor-conceived child was illegitimate and its mother guilty of adultery, even with consent from her husband. By law, pregnancy by artificial insemination with sperm from a donor was legally tantamount to an errant roll in the hay by a married woman with a man who was not her husband; it was adulterous, scandalous, and grounds for divorce.

Owen, my high school social studies teacher and mentor, occupied a prominent seat within my inner circle. He served as my chief intellectual ally whenever I needed his input. I always grew from hearing his perspective. During our discussions, he helped me intellectualize the social context of my discoveries as he asked me to recall my earlier studies in his class and make some connections. Eventually I could see my particular research interest within the larger social trends of the early 1920s: The Bolshevik Revolution swallowed Russia, eugenics advocates were spurred on by disclosures of donor-insemination, and the resulting rolling thunder of condemnation by church and state inspired fear and secrecy in those who needed help with conception.

This was the environment in which Aldous Huxley was inspired to concoct his satirical authoritarian society in which genetically engineered, selectively bred, perfect human beings were manufactured to create a perfect world order. *Brave New World*, his 1932 bestselling book, was required reading for every high school student.

Huxley's vision again seized the world stage after an explosive March 1934 magazine article titled "Babies by Scientific Selection." Designed to bring science to a narrow audience of armchair practitioners, *Scientific American* interviewed two hundred physicians located in seven cities across the eastern United States. A quarter of them reported patient requests for donor-insemination to accommodate a couple's male infertility. Fewer than 10 percent of them, eighteen to be exact, admitted attempting the procedure; nine claimed success in impregnating women with a sterile husband via the life-giving sperm from selected men. *Scientific American* declared it "one of the most significant eugenic developments in the history of man."[3]

The author, John Harvey Caldwell, concluded that from fifty to one hundred fifty test tube babies were born each year and that a growing segment of the population (one in ten couples) who were involuntarily childless could benefit from such treatment.

In a world where "our sterility is increasing" as we become "biologically weaker," the author painted a science fiction future scenario where "fertility clinics" in each city across multiple continents screened sperm donors as a piece of their practice to enable conception. They would meet first with resentment, then tolerance, then acceptance and, finally, with enthusiasm.[4]

Caldwell broadened his conjecture, building on Huxley's *Brave New World*, where humans were selectively bred to be genetically perfect. Caldwell's imagined "fertility clinics" would not only enable conception, but would also teach birth control and practice sterilization of the "unfit." Such clinics could even sponsor baby shows at county fairs, sponsor child health contests, and enable anyone to have a baby by "scientific selection."

Six weeks after the *Scientific American* article came out, in early May 1934, with the Great Depression in full swing and one in four people unemployed, the world's press latched on to this emerging trend and

---

3   John Harvey Caldwell, "Babies by Scientific Selection," *Scientific American*, Volume 150 (March 1934): 124.

4   Caldwell, "Babies by Scientific Selection," 124.

publicly dubbed children conceived via artificial insemination "test tube babies," or "laboratory babies." The press did not distinguish babies conceived via artificial insemination by husband from babies conceived via artificial insemination by donor.

A media frenzy of countless newspaper and magazine articles focused on the case of the Lauricellas. The Long Island, New York, couple had conceived twin girls by artificial insemination with sperm from the husband. The husband, a mechanic, had rented a parking space to the doctor and eventually relayed to her his eight-year story of the couple's failure to conceive a child.

Both the couple and the doctor took the unusual step of going public with their story and achieved their fifteen minutes of fame. The doctor stole the show, however, with a shattering revelation. Artificial insemination by husband was less controversial, by far, than artificial insemination by donor, which captured the attention of the press and the legal community. During the interview, the couple's physician, Dr. Frances Seymour, drew the public glare. As a woman physician and as a gynecologist, both her gender and specialty were unusual for the times.

Dr. Seymour launched a colossal story with her disclosure of two artificial insemination procedures she had successfully performed on two professional "prominent, unmarried women," using donated sperm from men she had selected from a blood donor list. The story garnered top media shelf space (front page, top right, cover story) in newspapers from coast to coast.

The press repeated *Scientific American*'s estimate of one hundred fifty children born per year who had been conceived via artificial insemination and its prophecy of eugenic-selected birth à la *Brave New World*. Was this living proof of breeding genetically perfect humans according to some sort of caste system that would define the children's place in society? Selective breeding had been practiced on crops and livestock for centuries. People, too?

Dr. Seymour's role as medical director for the National Research Foundation for Eugenic Alleviation of Sterility threw high-octane gasoline onto the fire of public controversy. The medical community at large felt hugely threatened by such a public disclosure and the resulting press fervor. Would politicians, religious institutions, academia's elite, shark lawyers, and society at large disparage the medical community? Were there legal threats, malpractice suits, or claims of adultery by doctor?

The New York Academy of Medicine, *JAMA*, and several other medical community advocates issued bold statements intended to dampen the press hype and public enthusiasm the very next day after the Lauricella story broke. They declared that artificial insemination was not new. Rather, since it was potentially dangerous and not proven to be all that effective, artificial insemination of any type was not a mainstream medical practice.

Yet, the public debate continued to rage. The media shouted claims of eugenics with a tone of righteous indignation. The medical community, and any interviewed doctor or spokesperson, stared back silently during press interviews as if adhering to a Mafia-like code of *omertà* (silence).

Newsweek joined the fray on May 12, 1934, less than two weeks after the Dr. Seymour story broke, with its own blockbuster artificial insemination article. Provocatively titled "Ghost Fathers: Children for the Childless," the piece highlighted the medical procedure that allowed a pregnancy to appear to be the result of marital sexual relations. A husband's infertility could be masked by this "ghost father," whose hair color, eye color, skin color, ethnic ancestry, and perhaps even religion and blood type were similar to the would-be father. He then faded away upon conception, a forgotten man, a ghost.

"Ghost father," I repeated to myself. My sperm donor was a ghost. I would likely never know who he was, what he looked like, his ethnicity, his traits, his blood type, and overall health. Yes, my dad was my dad, and I was a logical Italian, but what about my biology? My children did not seem to need this knowledge as much as I did. They were curious but did not want to express it for fear of hurting me. Both their grandmothers were prematurely widowed and had later remarried child-loving men. They had reminisced about their times spent with their step-grandfathers. Like me, they never met their biological grandfathers.

"What a void," I confided to Susan. "I want my eventual grandchildren to know their grandfather. I'll be no ghost."

Would-be parents needed to trust their doctor implicitly. The sperm donor's selection was the doctor's alone. The doctor guaranteed donor anonymity. Most doctors seldom kept these much-too-revealing records. When the process was executed perfectly, "Ah, he looks just like his father" would be the prevailing comment heard about the donor-conceived child.

Newsweek broadened the visibility and appeal of artificial insemination by donor by reporting that the Marriage Consultation Center in New York City would refer couples to doctors who were willing to perform donor-insemination, with a big IF—if the husband signed a document giving his consent to preserve the "mutual happiness" of the couple and the "well-being" of his wife. The center, and its co-founders, Dr. Hannah Stone and Dr. Abraham Stone, immediately captured the public spotlight.

The daughter of a pharmacist, Dr. Hannah Stone attended the first American Birth Control Conference in 1921, a year after receiving her MD from New York Medical College. There she met storied birth control and reproductive rights trailblazer Margaret Sanger. The two formed an immediate and passionate bond over the principle of enabling women to control their own bodies and reproduction.

Once Sanger opened the Clinical Research Bureau in 1923 (later renamed the Birth Control Clinical Research Bureau), Dr. Hannah

Stone first served on its medical advisory board and then joined the Bureau as a physician in 1925. She and Sanger, along with several colleagues, were harassed, arrested, and detained several times throughout their careers for disseminating birth control information and Dutch or Japanese imported diaphragms in violation of the Comstock Act.

Dr. Stone was largely credited with discovering the most effective birth control method of her day: the use of a diaphragm in conjunction with a newly developed spermicide. She went on to author an article, "Therapeutic Contraceptives," which was one of the first about birth control to be published in a medical journal.

Joined by her husband, Dr. Abraham Stone, the couple added relationship and sexual problems, including fertility and artificial insemination, to their birth control counseling. By 1928, they formally named their practice the Marriage Consultation Center, the first of its kind, which they ran from the clinic and within a community church.

Given their newfound notoriety, the Stones published a popular book for couples, *A Marriage Manual: A Practical Guide-book to Sex and Marriage*, in 1935. Written as a frank, factual, and intimate interview with an about-to-be married couple, it further amplified the donor-insemination process as an alternative for a childless couple to consider. Published by Simon & Schuster and acclaimed by both the medical and educational communities, the book established the standard for a practical guide to sex and marriage. Simon & Schuster subsequently updated the book and translated it into several languages.

I found the sociocultural history of artificial insemination by donor to be fascinating. It also helped me to imagine the conversations my parents had that ultimately led them to pulling off a hoax and gaining half a biological child.

My parents were conventional in many ways, but I knew both my parents to be creative thinkers, too. My mother made decisions while flying by the seat of her pants; she relied on her gut. My dad, on the other hand, was far more analytical. He would have researched the whole process more deeply before making a decision. They must have discussed adoption as one of their options. Did they turn away from adoption because of how things might be perceived? In 1945, an Italian man with a childless wife would be admitting that one (or both) of them had a fertility problem. Or was it more a concern about the child? If they adopted, what kind of child would they be taking into their lives?

These imagined conversations helped me to understand why they opted for the help of a hired man—an anonymous ghost.

Chapter 12

My research partner, Tracy, traveled home from New York City on occasion and we visited her grandmother. By 2004, my mother had moved into an assisted living facility. We sat in her tiny living room on a small sofa that faced where she sat propped on her favorite easy chair and enthusiastically told her about the 1934 "Ghost Fathers" article and asked her what she remembered.

"I was thirteen years old at that time," my mother said. "My father shipped out on a merchant ship, a freighter, to earn some money after his convenience store went out of business. We were alone in the middle of the Depression trying to survive."

She went on to add some closely guarded family history. "My mother, your great-grandmother," she hedged, looking directly at Tracy, "was trained as a nurse and midwife. She began performing abortions on the side for women who could not afford to have another child. I boiled the water. Then she kicked me out of the house until it was over."

Tracy observed out loud, "So women were quietly taking more control of their bodies and their lives even back then."

Her grandmother offered a smile and a nod. Intellectual or philosophical conversation was not her forte. She spoke in pragmatic terms, with a child-of-the-Depression-era perspective. "We had to eat. That paid for some of our food, I'm sure."

Tracy continued, "Do you think that maybe your mother and the women she helped influenced you in any way?"

I am a proud father for sure, but my daughter's intuitive ability to connect dots and the empathy she possessed when she connected them took my breath away.

"How do you mean?" my mother asked.

"Well, like them, you also took control of your life and your circumstances. And your husband? Really so progressive for the times."

My recollection of my healthy dad also took my breath away. He marched to the beat of his own drum in the way he lived his life, outside of the mold in which he was raised. He challenged conventional wisdom quietly and took calculated risks in his own style. He never made a big production about it; he just did it. *He would have adored my daughter*, I thought. She approached her own life in much the same way—unconventionally, with purpose, and never with a big production. She just did it.

My mother smiled, then laughed out loud. "Yes, he was a doozy," she said.

On many occasions thereafter, my family openly discussed the topic, either at our dining room table or around the fireplace. My sweet little old lady grandmother was an abortionist. During the Depression, back-alley abortions were often likened to a death sentence. Trained as a nurse, my grandmother offered a somewhat safer solution.

Birth control activists Margaret Sanger and Hannah Stone had been passionately saving lives by distributing birth control information and diaphragms to poor New York City women who could not afford the time and responsibility of another child. And at the same time, barren couples were desperate to conceive a child.

While the ensuing public discussion and debate enflamed controversy throughout the Great Depression, demand for artificial insemination services quietly increased. Yet, finding a doctor willing to perform the service remained challenging.

Many doctors considered the practice of artificial insemination distasteful or unethical. But throughout the 1930s, a small segment of doctors claimed "fertility" as their sub-specialty; thus, they established themselves as separate and apart from obstetrics and gynecology. As one piece of their practice, if insemination by the husband proved impossible, donor-insemination, sometimes termed "semi-adoption," increasingly became an attractive alternative to traditional adoption.

I had never given much thought to adoption before my mother's disclosure. I knew a handful of people who were adopted, as well as a couple who adopted a child due to the woman's uterine problems. Sperm banks had emerged well before my mother spilled her secret, but, again, I never thought about them. I imagined that people who could not conceive a child naturally, or via the help of a fertility doctor, for whatever the reason, would just adopt.

I felt shallow and self-absorbed once my mother revealed that I was the product of a sperm donor, not from a sperm bank, but likely from Harvard Medical School. My thoughts were all about me, without empathy for a young couple desperate to have a child and build a family. I floated the term "semi-adoption" to my mother.

She replied, "We never used the term. It's a good one, but your dad always considered you his child. And that's exactly what your birth certificate says." Her eyes shined with pride.

The medical and legal communities were becoming aware of the need for collaboration to address this new medical practice. As they were fully aware, malpractice suits could best be defended by expertly practicing "standards of care."

"Semi-adoption," I repeated to myself.

Faced with male sterility, my parents were given the option to "semi-adopt" and conceive a child that was biologically half theirs instead of raising someone else's child. I understood that option, but why keep that secret from me and create such a void? Was I a consolation prize?

Chapter 13

As I peeled the onion with my research, a culture of conspiracy and utter silence glaringly and unapologetically presented itself.

Prior to WWII, Dr. Seymour, who had worked with the Lauricellas, along with her husband, Dr. Alfred Koerner, who was also trained as an attorney, set the stage within the medical community. They worked behind the scenes, away from the view of the public and press. Their National Research Foundation for Eugenic Alleviation of Sterility, one of a number of Charles Davenport–inspired eugenics organizations founded after the Immigration Act of 1924, served as their initial platform. The foundation focused on helping the "right" types of parents conceive the "right" types of children.

In their article written for the *Literary Digest*, titled "Eugenic Babies," Seymour and Koerner featured their practice of relying on "middle-aged men with college degrees, professional standing, and a record of success 'from both a monetary and a scientific viewpoint.'"[5] They advised matching a donor's racial and social characteristics to the infertile husband's "so that a 'phlegmatic German' would not be bringing up a 'quick, fiery-tempered Italian youngster,'" and to prevent parents "who are both sandy-haired Scots" from being embarrassed by "presenting to the world a dark-eyed Spanish brunette." In their view, donor-insemination required a eugenics approach with the goal of creating a child with the desired perfect characteristics.[6]

Dr. Seymour went even further, requiring a minimum 120 IQ in all donor-recipient mothers. Even doctors who rejected those early eugenic goals generally first selected their "deserving couples," their patients, from a pool of White, native-born, better-educated, middle-class people.

Seymour and Koerner had developed a legion of followers. One early practitioner, Dr. R. T. Seashore from Minnesota, encouraged the procedure only for those who were likely to make positive contributions to society. Another, Dr. Grant Beardsley from Oregon, confided to Dr. Seymour that he had selected couples with high moral standards who were highly intelligent and financially capable of providing a child with educational opportunities befitting their social standing.[7]

Attaining some degree of 1938 "proxy father" notoriety was Dr. Ivy Albert Pelzman, who founded a sperm donation center within the Georgetown University School of Medicine. He grouped his medical school student donors (he paid $25 for each donation) by background and ethnic and physical characteristics. Dr. Pelzman proudly presented

---

5   Cynthia R. Daniels and Janet Golden, "Procreative compounds: popular eugenics, artificial insemination and the rise of the American sperm banking industry," *Journal of Social History* (September 2004): 9.

6   Daniels and Golden, "Procreative compounds," 9–11.

7   Daniels and Golden, "Procreative compounds," 9.

as an example his Chicago patient who had two donor-conceived children who "looked just like their father."

In June 1941, just six months prior to Japan's attack on Pearl Harbor and the subsequent decision of the United States to join WWII, Seymour and Koerner released their survey of 30,000 doctors regarding artificial insemination. *JAMA* published the results. One-quarter of the doctors had responded, with half of the respondents claiming "personal knowledge" of the use of artificial insemination. The survey identified one hundred practitioners, in every region of the country, who defined their practice as "fertility specialist."

Based on the survey evidence, Seymour and Koerner estimated that artificial insemination accounted for as many as 10,000 births, with approximately two-thirds resulting from artificial insemination by husband and one-third from artificial insemination by donor.

One-third by artificial insemination by donor—over 3,300 donor-conceived babies! That estimate stunned the medical community, as well as followers outside medicine—this estimate was multiple times larger than the size of *Scientific American*'s estimate of perhaps fewer than 1,000 births in total, mostly attributed to artificial insemination by husband.

In an environment in which doctors and patients alike were clandestine, professional skepticism prevailed. While the medical community argued about the survey's accuracy, it fully acknowledged that artificial insemination had grown in practice and in reliability. The "hired man," while not fully embraced by much of the medical community, had gained credibility among a small, innovative group of practitioners as the most effective treatment available for involuntarily childless couples with an infertile male.

While doctors and patients alike navigated the pre-WWII tainted social status of artificial insemination, standard methodologies began to emerge and influence the manner in which doctors talked with and treated their patients. The standard practices provided all parties with adequate legal cover.

My history studies reminded me that for every dogma that gets established, a schism follows. I had discussed this with Owen: "Christianity had several. It looks like so did the artificial insemination by donor handbook," I told him.

An unknown but sizable portion of fertility practitioners had added what my attorney friends termed "representation fraud" to their code of secrecy to enable an infertile husband's psychological paternity of a donor-conceived child. First, the fertility specialist might have instructed a couple to return home and have intercourse just after the wife received the donor's sperm, having claimed the process had intensely boosted the wife's fertility. Second, the doctor might have mixed a husband's impotent sperm with a donor's fertile sperm, claiming it boosted the husband's sperm and made it more potent.

The wheels of progress continued to advance, even as the ravages of WWII took center stage in lives everywhere. Founded in 1944, when most of the world was focused on the Allied invasion of Normandy, the American Society for the Study of Fertility and Sterility published a journal, *Fertility and Sterility*. The Society held regularly scheduled meetings and discussed perplexing issues such as the timing of ovulation and the viability of sperm.

Fertility doctors' practices were shaped by the fear of the social reproach, even disgrace ("social opprobrium" was the term they used), of participating in the birth of a "test tube baby." The advice provided in the Seymour and Koerner case studies and by a small subset of fertility practitioners formed the basis of fertility practice standards when dealing with a sperm donor.

Standard methodology, or "best practices," became another term for "standard of care" guidelines for artificial insemination by donor. The guidelines gave the fertility doctor legal shelter from any malpractice claims and provided a solid foundation for successful artificial insemination by donor.

First and foremost, a doctor needed to identify a "deserving and exceptional couple in a stable, permanent marriage, whose marriage would not collapse with the daily confrontation of a semi-adopted child." Could this couple hold a lifelong secret?

It might be easier to gain admission into Harvard Medical School, I thought.

How had my parents, with my dad's four-syllable last name containing five vowels, passed the admissions test? Had I been on the other side of that table, I would have seen a devoted couple who had been married for five years, with a strong work ethic. The wife was attractive, brunette, and vivacious. The husband, with thick, brown, wavy hair and crystal-blue eyes, was handsome, physically fit, gainfully employed, technically educated (although not a college graduate), and accomplished in his own right. I would have seen a "deserving and exceptional couple." Whether they could keep the secret would have been a wild card, but I would have placed the bet.

Initial treatment had to begin with extensive fertility tests of both the woman and the man. The test of the husband's fertility needed to stand up in court if, for some reason, after a donor-inseminated pregnancy, the wife conceived naturally.

Secrecy prevailed as *the* practice standard; only the doctor, the female patient, and the patient's husband should know that a sperm donor had enabled conception. The principal goal was to emotionally protect the new family from society's condemnation and the damaging gossip that would likely ensue. Would the woman be labeled an adulteress, or the doctor an adulterer? How about a conspirator? What about the pride or the questioned masculinity of an infertile husband? Would the couple, child, or donor be targets for blackmail? Is it grounds for divorce? What about child support in the event of divorce?

And, of course, there was a prevailing concern about any possible damage to a child's psyche. Seymour and Koerner wrote in their widely read article, "The Medicolegal Aspects of Artificial Insemination," "If a donor child were to learn of his or her origins, an inferiority complex would be set up with a root that psychoanalysis could not destroy and the child's maladjustment to society would result."[8]

In other words, this secret should last a lifetime, not only for the life of the couple and the donor, but also for the life of their child. Family law was completely ill-equipped to deal with the ramifications of the existence of a donor-conceived child on inheritance law and any and all of the legal ramifications of anyone complicit in his or her conception.

As my mother's secret unfolded, I wondered why my parents did not use sperm from my dad's close nephew, Frankie, or his closest friend, Ray. The fertility doctor's WWII practice standards provided my answer. If a couple wanted to use sperm donated by a family member or friend, fertility practitioners were advised to outright refuse. Central to the argument was an age-old concern that a wife's affections over time might transfer to the donor. Family and friendship circles would just complicate an already complicated situation.

My mindset was so twenty-first century.

As anyone who holds a secret can attest, the larger the circle of people who know the secret, the larger the chance that a secret will be leaked. Koerner's fertility practice standards advised keeping the mother's obstetrician, who would deliver the child, separate from and unaware of her fertility doctor.

The donor needed to provide his sperm at another location or at another time to prevent an unintended meeting of the couple and the donor. To make this conspiracy uncompromisingly legal, the mother's

8   Frances I. Seymour and Alfred Koerner, "Medicolegal Aspect of Artificial Insemination," *Journal of the American Medical Association*, Volume 107, Number 19 (1936).

doctor or obstetrician or midwife needed to sign the child's birth certificate without reservation indicating that the husband was, indeed, the child's father. If the semi-adoptive father sought legal adoption, the conspiracy, and the secret, would surface. It would require the donor to be identified and to relinquish all rights and privileges afforded a parent and be held harmless for child support. The child's inherited assets would, thus, be protected. The prevailing view was that formal adoption simply added enormous complexity to the circle of secrecy. *Omertà!*

My daughter and I drew blanks at the outset of our research. No wonder!

Fertility practitioners operated legally, but in the darkest of shadows. They maintained a stealthy profile because of an uncertain social acceptance. That stealth sure helped explain our lack of initial research success at the Boston Public Library.

We found no listings of 1944–1945 doctors in the Boston area that even remotely identified themselves as fertility specialists; not even a gynecologist had an office on 10 Beacon Street. Fertility doctors practiced invisibly and remained the sole gatekeepers in the selection of patients and donors in much the same way as doctors were gatekeepers for abortions.

By the mid-twentieth century, most states had criminalized abortions, with the exception of the therapeutic abortion. The doctor remained the reigning authority over life or death in the case of either unwanted pregnancy or involuntary childlessness.

The illegality of abortion, state by state, was clearly defined in varying courts by WWII. Elective abortion was not an option for patients. Only a doctor had the discretion to make that life-or-death decision, which solely took into consideration the life of the mother. In several states, it was also illegal for a doctor to provide contraceptives to patients. It was generally thought, at least in the medicolegal community, that the artificial insemination of a woman, using her husband's sperm, did not present the doctor any legal concern, although some in society held preconceived notions and religious objections. But donor-insemination caused legal heartburn.

Dr. Koerner, the doctor/lawyer, provided early fertility practitioners with a set of legal forms to both protect themselves with legal cover and establish legitimate children in the eyes of the law. His procedure required a couple to present their marriage license, notarize their signatures, and provide fingerprints on their consent forms. This also prevented a woman from posing with a male friend as a couple to gain access to a donor's sperm without her husband's consent. He included a standard release of liability that disqualified the couple from future malpractice claims, and a consent form for both the donor and the donor's wife to protect the donor from future claims of adultery should his marriage later dissolve.

Finally, Koerner advised a system of secrecy that included a figurative lock and key for all the documentation. His article on the entire system of secrecy and all the suggested legal forms, published and reprinted in medical literature, were made available to all fertility specialists.

I reflected on the several conversations I had had with my mother since she had unfolded this conception secret. In 1941, as Franklin D. Roosevelt declared war on the Empire of Japan for its December 7th attack on Pearl Harbor, my mom and dad were planning their family. After being married for two years, they sat around their radio in their Newport, Rhode Island, living room after the president's radio address.

I imagined their conversation: "In what kind of world will we be raising our children in?" They had no idea at the time that they would be scheming to have a child in such a clandestine, unorthodox manner.

By the end of 1941, despite Dr. Koerner's published approach and legal review, *JAMA* had editorialized that donor-conceived children were flat-out illegitimate. If a man knew he was not the child's biological father, whether by way of adultery or a procedure in a doctor's office, his only option was legal adoption. A few other physicians practicing artificial insemination by donor had editorialized that a more "common sense" approach would be best; zero documentation and no permanent file to corroborate any legal claim of "wrongdoing."

Artificial insemination by donor was the topic of editorials and case studies in various medical journals in the early 1940s. And after publication of the Seymour and Koerner survey, the legal community woke up to the pressing need for discussion, debate, and resolution of the legal issues surrounding artificial insemination by donor. The study had revealed that several thousand people carried a disputed legal status of "illegitimate" or "bastard," in spite of standard medical practices and the notion of "semi-adoption." Whatever the number, their ranks were exponentially increasing.

In early 1945, the legal and medical communities came together as the Chicago Symposium to discuss, among other matters, legalities of artificial insemination by donor. Neither the legal nor medical authorities coalesced into a single prevailing point of view. For every practitioner advocating for acceptance and legal change, detractors argued against it. The only actual agreement was on the need to establish a committee to study the need for and type of legislative changes required to address unresolved legal issues that impacted the practitioners and the resulting children born by such controversial means. The issues were both medicolegal and sociolegal, with no ready or easy resolutions.

In anticipation of this Chicago Symposium, on February 26, 1945, *Time* published an article in the form of a case study on a lawsuit in the Circuit Court of Cook County, Illinois, *Hoch v. Hoch*. This suit granted a divorce to the husband on the grounds of adultery via artificial insemination by donor and questioned the legal status of donor-conceived children. It carried the dubious title "Artificial Bastards?" Yes, with a question mark.

The popular press engaged social and political leaders to continue the debate. Should the donor be anonymous or non-anonymous? When should the children conceived via artificial insemination by donor be told about their biology—and should they be told at all? Is it possible or advisable to use sperm from a relative (such as a father or brother)? Should the donor be compensated? How much?

After I found this old article in early 2005, I talked with my mother, who was nearing the end of her life. "Do you recall a *Time* magazine article from around the time that you and Dad were in the midst of the donor-insemination procedure with your doctor?" I gave her more details.

She smiled with her customary twinkle in her eyes. "We didn't read *Time*. We read *Life*."

I knew that to be true. The local newspaper and *Life* were staples in my childhood home. Publications like *Time* and the *New York Times* were not. Even though my mother had come clean with her secret, I had a lingering sense that she had been more informed than she was letting on.

(Over a decade later, I would find evidence that the popular press of the 1950s and 1960s heightened the general public's awareness of families in my parents' shoes who weren't about to settle for barrenness as an "act of God." *Time, Newsweek, Collier's, Woman's Home Companion, Coronet, Good Housekeeping*, and *Ladies' Home Journal* had joined the fray with a litany of provocative titles: "Proxy Baby," "Born to Order," "The Riddle of AI," "Test Tube Babies: The Controversy over Artificial Insemination," "Secret of AI," "Frozen Fatherhood," "The Child of Artificial Insemination," "Artificial Adultery," "Test Tube Baby: A Woman's Right?," "Test Tube Babies for Single Women," and "Are These the Most Loved Children?" I just knew deep down inside that my mother had devoured a host of those articles. I recalled seeing *Ladies' Home Journal* and *Good Housekeeping* scattered about on the kitchen table or the living room sofa. Why hadn't one or more of those articles made her come clean to me? Was it just easier to ride with the secret?)

This debate raged in not only the United States, but also throughout western Europe. In England, in post-war 1945, Dr. Mary Barton documented in *The British Medical Journal* what she labeled as successful artificial insemination by donor and later revealed as many as 1,500 donor-conceived children in the United Kingdom.

The reaction to Dr. Barton's article further unraveled the cloak of secrecy around the practice of artificial insemination. She was vilified by the press and condemned worldwide. The outrage cracked the Richter scale when she subsequently disclosed that one anonymous donor, her husband and fellow fertility researcher Bertold Wiesner, accounted for six hundred of those children. (Wiesner occupies the number-one Wikipedia spot for having fathered the most children.)

The Archbishop of Canterbury and the Chief Rabbi of Orthodox Judaism had condemned the practice of artificial insemination by donor along the same lines as the Roman Catholic Church. Other religious leaders throughout Europe followed suit. Separating biological and social paternity flew in the face of centuries of tradition used by church and state to establish clear lines of paternity. Introducing a third party between husband and wife would not see the light of day in other cultures.

The issue triggered ancient anxieties among men in patrilineal societies that they might be deceived by women into claiming another man's child as their own. They felt that if sexual intercourse between a married woman and a man who was not her husband was adulterous, so, too, was pregnancy by a donor in any form, albeit artificial. The church and the law were in agreement.

Amid this post-WWII noise from the courts and religious, political, and social leaders, the experts repeatedly estimated a constituency of involuntarily childless couples to be 10 to 15 percent of all couples; this number was reported and re-reported in medical journals.

Female fertility issues were estimated to have accounted for 40 percent of the infertile couples. Reproductive scientists reported that a woman's fertility could be impacted by several factors: ovulary, tubular, cervical, uterine, hormonal, weight, over-exercise, age, or some underlying medical condition. Researchers stated that a combination of male and female fertility issues had compounded a couple's conception difficulty in 20 percent of the cases.

Studies concluded that 40 percent of the couples were childless due to a sterile male. Male fertility could be impacted by testicular, glandular, or hormonal factors, which caused sperm dysfunction in either number or function, notwithstanding any underlying medical condition; to a far lesser degree, male fertility could also be affected by age.

My dad was forty-one years old when I was conceived. His underlying medical condition that most likely led to his sterility was the trauma of those twenty-six operations and morphine use during his teenage years—the same condition to which I had fictitiously attributed his struggles with depression.

A growing cadre of medical specialists had the expertise, the vision, and the willingness to provide that constituency of involuntarily childless couples with a service; it was a silent service and not openly promoted due to the sociological stigma. Those childless couples had two simple choices: pregnancy via a "hired man" (the sperm donor) or adoption. The businessman in me said, "Wow! That was a sizable, underserved, niche market driven by consumer demand!"

In my technology business dealings, I learned that the innovators in any emerging market, affectionately termed "the Lunatic Fringe," prove the viability of a new technology. They compose the first 1.5 percent of any emerging market. Once the technology is proven, the next 13.5 percent of an emerging market, the Early Adopters, add their testimonials and increase the marketability of the innovation. That enables the more risk-adverse Early and Late Majority of a market to adopt the innovation. Lastly, the Laggards, the final 12 percent of any emerging market, provide the final leg of growth to a maturing market that has already experienced its big spike in growth.

In the aftermath of the worldwide knowledge and horror of Hitler's Holocaust, eugenics advocates all but disappeared from the public debate. As the artificial insemination debate raged, amid worldwide condemnation and de-legitimization, and just two weeks or so after *Time*'s "Artificial Bastards?" article, two Early Adopters, my mother and the man I knew to be my father, made a decision. In a process some labeled "adultery by doctor," they took the steps—with the help of a "hired man," a ghost father—that led to my conception. They were part of an ever-so-tiny, courageous group of childless couples willing to challenge social convention to have a family. They enlisted the help of another Early Adopter, a fertility specialist (the undiscovered Dr. Sims?), who practiced without fanfare.

Despite my mother's declining health, she was able to remember and share more and more information when I asked her some new and detailed questions that were the result of my research.

"Yes, we signed a bunch of forms," she explained, "but I really don't remember what they all were. Likely legal stuff for the doctor."

When I asked her to further recount the *omertà*, she commented in matter-of-fact terms, "Your dad was an Italian Catholic."

I could interpret what she meant. My dad was not devoutly religious, but his immigrant family was not open to extending the boundaries of natural law in such an avant-garde manner.

Tracy commented on some of her grandmother's family lore during one of our Sunday afternoon phone calls. "She was raised as an Episcopalian only because her family lived next door to the church," she told me. "Her parents weren't the least bit religious either."

I agreed. "The church next door was good enough, and certainly convenient enough, to give their children some spiritual exposure. She converted to Catholicism as an adult so that she and her husband could take their wedding vows at the altar of a Roman Catholic Church."

All for appearances, I thought.

My dad was not marrying an Italian, but at least she was Catholic.

My first-generation, early twentieth-century American dad was breaking the family mold yet again. His immigrant family had practiced a classic form of endogamy. Italian Catholics from the same little village near Bologna married neighbors. They lived in a closed, safe society. When they immigrated to America, they congregated into cohesive neighborhoods and married others from the Bolognese region.

Endogamy was not unique to Italian Americans; it was also prevalent within the Pilgrim colonies and Colonial American communities. American mobility and the melting pot of the twentieth century had changed the endogamy dynamics of love and marriage.

I smiled at that realization. My dad, once again, was an innovator—part of the Lunatic Fringe.

"Never mind society," my mother continued. "What about his family? We protected him, and we protected you." She added, "Our doctor insisted on secrecy for life. I violated that pledge when I told your grandmother and swore her to secrecy, too. Your legal adoption would have blown our closely kept secret wide open."

I silently wondered how my father's family would have reacted to their infertile hero. Would they have pointed fingers? Would they have abandoned their hero in the same fashion as they did when my dad was debilitated by depression years later?

My parents opted for an administrative solution to a technical problem. The whole process was well engineered according to the "handbook"; it was a successful collusion.

The doctor who delivered me was totally unaware of my mother's visits to a fertility doctor and named her husband as my father on my birth certificate. I am legal and legitimate. I have a father's family name, his Northern Italian blue eyes, and normal maladjustments that have nothing to do with his mental illness, suicide, or my later-in-life knowledge of my donor-paternity.

On another of our catch-up phone calls, Tracy and I talked about how the fields of gynecology, obstetrics, and fertility were male-dominated at the time of my conception. Tracy intuitively observed, "Gram's mother was her role model. They both took back the power of their female body."

I agreed. My Scotland-born grandmother moved at odds with society's norms in her own fashion. Her eyes would often tear up when she looked at me and said, with a glow, "Miracle Child." I felt desperately wanted and unequivocally loved in one side of my head, and depressingly angry at the violation and deceit in the other side of it.

Chapter 14

During my many visits to both Harvard's Medical School library and the Boston Public Library, I struck up a friendship of sorts with Liz, an enthusiastic Boston Public Library volunteer. A wonderfully congenial former grade school teacher with puffy gray hair, she reminded me of Mrs. Cook, my very sweet first grade teacher. As I continued to research the history and sociology of artificial insemination, I confided to Liz that the circumstances of my conception—artificial insemination by donor—served as my motivation for all my research. I asked, "Could you help me find others like me, who were donor-conceived?"

She might have hugged me; instead, she adjusted her oversized pink-framed glasses and said warmly, "I'll see what I can do."

Perhaps two weeks passed before I visited the Boston Public Library once again. It was a dreadfully cool and rainy Saturday morning in the early spring of 1999. Over three years had passed since my mother's revelation. Liz was beaming at the front desk as I approached. "I've been waiting for you. I found a few things. Some you might not like, but here they are."

"You're so kind. Thanks so much, Liz," I said as I took the oversized manila envelope she passed to me. I sat at a vacant table in a darkened corner of the library and opened it. Within a yellow folder in the envelope Liz had given to me were four Xeroxed articles. At the top was a 1955 *New York Post* piece reporting that 50,000 donor-conceived babies had been born in the United States between the onset of WWII and 1955. Their number was forecasted to grow by 6,000 children per year. The *New York Post* was not on my reading list as a grade schooler; it had never been on my reading list. As an adult, I read the *Wall Street Journal* daily and sometimes on Sundays I read the *New York Times* rather than the *Boston Globe*.

The next item in the folder was a Sunday, April 18, 1976 *New York Times* article by Lillian Atallah titled "Parent and Child." She told her story about her loving parents who disclosed to her at the age of nineteen that she was donor-conceived, with the aid of (oh my gosh!) Dr. Frances Seymour.

As the author was growing up, her parents had taken her to Dr. Seymour for periodic "checkups." She had felt as though the doctor had behaved more like a gleeful aunt, with praise on how well she had developed physically, intellectually, and emotionally. Once Atallah realized the context of those visits, it all made sense to her. And her experience resonated with me—I was glad to know that someone like me, a donor-conceived person, could be happy and well-adjusted.

The third item in the folder was a *New York Times Magazine* article titled "A Scandal of Artificial Insemination," by Judith Gaines, dated Sunday, October 7, 1990. Gaines first cited reproductive industry statistics: over 80,000 women artificially inseminated by anonymous donors yearly by 11,000 practicing physicians, with the aid of 400 sperm banks producing $164 million in annual revenues, yielding 30,000 annual births.

Those statistics were not scandalous on their own, but survey results from the Congressional Office of Technology set off alarm bells in my head. More than half of those practicing physicians surveyed performed no tests on donating men or their semen—not for AIDS, not for venereal disease, not for hepatitis, not for genetic disorders. Sperm banks were found to do a better job of testing for AIDS but generally failed to test for anything else.

With no industry regulation, all of the individuals involved were playing Russian roulette with human life. How many catastrophes had already happened? How many more were yet to happen to the unborn? Had Dr. Sims thoroughly screened my parent's donor? I was healthy. Perhaps Dr. Sims had done a comprehensive screening . . . or perhaps I had won at Russian roulette without ever realizing it.

The last article in the folder was the longest; it was also a *New York Times Magazine* article, published on Father's Day, 1995, titled "Looking for a Donor to Call Dad," by Peggy Orenstein. This one was published five months before I learned I had been donor-conceived. The article detailed the experience of two donor-conceived people who discovered their family's "dirty little secret." One was a man, conceived in 1949, Bill C., who confronted his mother after the death of his stern, unaffectionate dad. She finally told him the truth. After years of searching and learning that he had no rights to crack the secrecy surrounding his genetic origin, he doggedly tracked down his parents' physician, whom he suspected was his birth father. When confronted, the doctor admitted nothing and cited something biblical. Bill C.'s continuing investigation had uncovered that the doctor was, in fact, the anonymous sperm donor. Bill had dozens of half-siblings, all from the same general community, most of whom had no idea they had been donor-conceived.

Sitting at that darkened table in the Boston Public Library, I gasped. How might I have felt if I had uncovered that the mysterious Dr. Sims had perpetrated a fraud on my parents and acted as their anonymous sperm donor? And moreover, that he had done so not only for them, but for dozens of other unsuspecting patients? I would feel criminally violated—and that was just for starters.

As I left the Boston Public Library, I waited for Liz to finish a conversation with a library patron. With tears in my eyes, I hugged her and said, "You are the best."

Liz hugged me back as she whispered, "Good luck."

Back in the den of my Boston condominium, I typed into my computer the crazy name of an innovative start-up company in Silicon Valley—Google. I asked Google how many babies had been born in the United States since 1940. Almost instantly, Google presented a table of registered births, by year. After some quick addition, I calculated that approximately fifty million babies had been born in the United States during that time frame. From that number, I calculated that children conceived via artificial insemination by donor in my age group in the United States represented an infinitesimal 0.001 percent of that total.

The medical community estimated that by 2010, one million people had been conceived using artificial insemination by donor throughout the western world. Most of them, especially the older ones like me, were, and remain, unaware of their rarity. With seven billion people occupying the planet, that is rare; 0.00014 percent is exceptionally rare! Fill up a World Cup stadium for the big game ten times over and you might find one like me.

My conception's timing predated sperm banks by more than thirty years, but I could not help but wonder just how special I was. How many other childless couples had been given access to that same sperm? How many times did my ghost father donate? Did I have half-siblings? Would we have anything in common outside of our DNA? Was it possible that I had dated one of them (oh, please NO) in college or later? Am I a child of Dr. Sims?

Whether my ghost father was my parents' doctor or a medical student at the time, he was likely a doctor.

I floated the idea by my own physician, Harold.

"It's certainly possible that the doctor was unethical. It's far more likely that your donor was a med student," he said, "but your penmanship is much better than any doctor's." He smiled. "Perhaps you've inherited a recessive gene!"

Upon learning all of this about my origins, I often joked with my closest circle of friends, "My father was a doctor; how about yours?"

But what about my paternal seed, really? Was he a Harvard-trained physician? How would I ever discover which one of the names of the 1945–1948 Harvard Medical School graduates in that yellow folder tucked away in the back of my briefcase was my ghost father, if his name were in that folder at all? Never doubt my resolve.

PART III:
THE REVELATION

Chapter 15

Discovering all of this about my origins, about the deep and sometimes sordid history of how I came to be, started to change me.
My insides felt as though they had been turned inside-out, draped on a
man-sized saguaro cactus, exposed for exhibition in an invitation-only
gallery, cured in the scorching Sonora desert summer sun, smeared with
a large jar of healing Vaseline, and then turned outside-in once again.

My persona had become softer, yet I had grown stronger, both person-
ally and professionally, as a result of my intense, identity-challenging ten
years from 1995 to 2005. The unfolding closely held secret of my conception, a marriage near the brink, deeply penetrating marital and individual
therapy to treat old wounds, years of fruitless searching, and researching
to find the truth behind my genetics combined to create stressful and
game-changing situations.

Gaining this deeper understanding of where I came from and who I
am went a long way in helping to heal my marriage. Therapy hammered

into me that "flawed" is the human condition. It was okay to be blemished. "Know thyself," said the Greek philosophers. Having fears and anxieties or feeling vulnerable were all part of being "normal." I had sixty years of practice at hiding those feelings. I had recently learned to reveal more of myself within my inner circle and to connect more intimately with my wife and others close to me. Perfect? Not by any measure. A huge self-improvement? A big yes! Popeye the Sailor Man had been my ally throughout this journey. I heard him repeatedly: "I yam what I yam!" I added, "I'm better than I yam."

The recessionary economic dynamics in the wake of 9/11, along with the bursting of the internet bubble and its aftermath, added acute professional drama to my personal journey. Real life continued to happen around me throughout my research and therapy.

My once-distressed company was no longer a shipwreck and had turned a profit. With its stock value having appreciated nicely, the venture capitalists who had previously occupied the firm's board of directors moved out and installed a new board. Technology is never static. Formidable firms had entered the competitive field. I advocated selling the company to an interested acquirer and distributing any gains to shareholders. This new board had visions of building the company for the next decade and dismissed me. Inadequate? I did not think so and had nothing more to prove. I took my own profits and moved on with my life.

While encouraging my son to apply his creative talent to the digital working world, I was also available to drive full-throttle into the internet bubble. I went into it headfirst as CEO of a dotcom that enjoyed a 10× IPO pop before the bubble burst. I was then head of a distressed organization once again, repositioning to survive and negotiating its acquisition by another public company.

In the middle of it all, the September 11 attack and its resulting war drums provided the catalysts for rekindled jungle flashbacks and resentment that I had kept hidden and private for four decades.

Expressing that resentment was another matter. Prior to 9/11, I rarely discussed my combat experience; no one wanted to know about it. After the World Trade Center attack, special operations was the topic of every news broadcast. It seemed everyone I encountered wanted to know. Still, I was not forthcoming. Old reservations behaved like rusty chains that needed a hard tug before releasing their grip.

Like so many other Vietnam veterans who had remained silent upon their return home, I began hearing, "Thank you for your service" for the first time. America, it seemed to me, had recognized its need to make amends for the hostility it had directed at homecoming GIs, hostility that had really been meant for its political leadership.

I have never been a political protester or demonstrator; I have too much baggage from my time in uniform. I swallowed in distaste as my family joined political protesters and marched in New York City and Washington, DC; I was supportive of their message, not their method.

Wearing an American flag pin upside down on my suit lapel at various business and social events both prior to and after the 2003 Iraqi invasion was as close as I had come to taking a public stand. I was quite surprised to learn how few people understood flag etiquette. An upside-down flag symbolizes an entity in distress. I explained this distress signal to anyone who attempted to right-side-up my upside-down flag pin.

On occasion, those interchanges led to some discussion of the war, both pro and con. I shared my views, which were bundled with some personal experiences of which most people were not aware and certainly did not possess themselves.

While sharing holiday cheer at a business event with two thirty-year-old investment bankers who had advocated for a retaliatory attack on what I had considered to be the wrong target, my blunt words flowed.

"Young men died in my arms over the lies of a president. Different time, different president, different political party, different lies, but dead kids, nonetheless. Upon what personal experience do you base your views?"

These bright, well-educated, pedigreed young men were speechless as they wore their life's inexperience on their sleeves and their faces. I loved that blank, stuttering look.

"We never knew," commented one about my experience in war.

"We never met anyone who had been in combat before," said the other.

The old chains had finally broken. I felt released from a long-term, self-induced confinement and was forever emboldened thereafter to reveal the secret caverns within myself.

Chapter 16

A handful of consulting projects that I had engaged in after the dot-com implosion provided me an opportunity to weigh my options. I could see a pathway to either my early retirement at age fifty-eight or a career change. Life never stands still.

In early 2005, a full ten years after my mother's secret first came to light, she died of congestive heart failure caused by that second leaky heart valve.

Throughout my formative years, it had been just the two of us dealing with the sickness and untimely death of my dad. Of course, I knew his sickness was not her fault, but her scattershot response to that distress, which had created such day-to-day chaos and change, had angered me well into my adulthood and independence. That anger impeded our relationship's harmony.

It took me a decade of adult perspective to understand that she deserved a gold star. When I was faced with a challenge, she'd often say,

"Just do the best that you can." And she did the best that she could within her unfortunate circumstances. She provided as best she could for her son. She set an example of tenacity and grit with her own sense of humor and grace. Was the origin of my own tenacity a piece of my maternal or paternal DNA, or had I just simply learned it from her example?

My former adolescent anger softened as I grew to recognize that I, too, was an imperfect adult who had made his own share of mistakes that inhibited deeper intimacy in my relationships. My past work-related separate living arrangement had increased the emotional distance in my already strained marital relationship. For a time, I was home on weekends only, with a few exceptions. My driving motivation (based in my childhood experience) was to maintain stability for my daughter. By this time, my son was a college student who visited home for holidays and summers. He was preparing to embark upon his independent life. But my daughter was still in high school and I wanted to avoid the potential volatility of moving her from school to school. Providing a stable environment for her was the right motivation, but I was going about it the wrong way! Susan and I worked feverishly in marital therapy to get past our estrangement. We were finally in a good place.

With my mother's death, the knowledge of who I was, of my artificial origin, had been swept away. The last known living person who could connect me to my missing gene pool was now forever gone. Tracy and I had just lost our sounding board for whatever we might have uncovered next. My mother's passing magnified the emptiness that continued to grow within me, an emptiness caused by the lack of genetic information about half of myself.

My knowledge of who I was, my genuine origin, had been reinvented. I felt that the secrets of my origin had forced me to live a compounded lie. I repeated to myself, with my customary resolve, "No more secrets!"

Had my mother lived even several months longer, it is doubtful that the next chapter of my life would have evolved quite the way it did. More likely, I would have stayed in Boston, which was the East Coast epicenter of venture capital (VC). I was exploring venture capital for what I thought might be my last career move.

The ranks of people working in VC were filled with silver spoons who held top educational pedigrees, or so it seemed to me. Why couldn't I join those ranks, too? Far too few VCs had operating experience at the CEO level. Quoting Shakespeare became part of my repertoire. "When the sea was calm, all boats alike show'd mastership at floating" (*Coriolanus*, Act IV). I figured that a tested ship (me) had earned a seat at the VC table, and I began exploring opportunities.

In years past, I had needed success to feel validated and respected and avoid feeling flawed and abandoned. I had grown, and I knew I was good at what I did. Few competitors had the stomach for the types of dysfunctional situations in which I excelled. I enjoyed the hot water leadership challenge, not for anyone else, but for me. I felt confident that I could now take a VC challenge and advance myself several steps further while teaching others, not just as the head of a technology company with issues, but as the coach of a number of companies. Being a mercenary at heart and repairing troubled entities and positioning them for valuable exits was a highly lucrative undertaking if it succeeded. I wanted to take that a step further as a VC as the grand finale to my executive career.

The transition from operating executive to venture capitalist worked! As the Go-to Guy for organizations in distress, an executive search firm approached me about joining a fifty-plus-year-old former high-flyer, a New York Stock Exchange–listed VC holding company in hot water both during and after the burst of the internet bubble. Its stock had cratered in spectacular fashion to pennies per share, a 99 percent loser. The firm was on the edge of bankruptcy and delisting from the stock exchange. It was not searching for a venture capitalist. It needed a tested ship, a CEO to rescue a catastrophic shipwreck.

The competition for a tough assignment like that was always thin; it was a perfect fit for a scarred and tested spoon not made of silver. And it was in a different city: Philadelphia instead of Boston. Philadelphia was out of the VC geographical mainstream; it was not Silicon Valley, not Route 128. Susan thought the role was a great fit for me, and for her. She was excited to be closer to her Pennsylvania family and aging mother.

After performing some due diligence, I had confidence that a few quick, bold strategic moves could turn this assignment into an equally quick success. Little did I fathom at the end of 2005 that this role coincided with the acceleration of DNA technology and would give me an insider's view of the new innovations that would help me discover my genetic origins.

My executive business experience was primarily in the information technology arena: hardware, software, and information services. This VC experience introduced me to the life sciences field—diagnostics, medical devices, and specialty pharmaceuticals—and further exposed me to a growing trend in healthcare called "personalized medicine."

In collective wisdom fashion, members of my firm's advisory board gave me an education. I learned that medical treatments could be more effective if they were customized based upon an individual's unique molecular composition. DNA technology had advanced exponentially. Science and computerized analytics converged in the life sciences community to breed new cost-effective and speedy analysis of an individual's chromosomal makeup.

In a further advance, that analysis had become available to consumers over the internet. My staff enthusiastically discussed 23andme (creatively named for the twenty-three pairs of chromosomes in a human cell), a Silicon Valley start-up established in 2006 to analyze an individual's chromosomes. Its high-profile co-founder was married to one of Google's co-founders at the time. The company's market introduction of the industry's first consumer DNA test over the internet in November 2007 was met with great customer enthusiasm and global fanfare.

With consumer access to DNA testing, new stories began to appear in the media about DNA discoveries by adopted and donor-conceived people (most of the latter had a sperm bank origin). Both camps had continued a practice of *omertà* into the new millennium, and information about conception via artificial insemination was disclosed to very few children.

The first reported DNA disclosure of artificial insemination by donor was a fifteen-year-old brilliant young man who had just uncovered his donor, formerly an anonymous sperm bank ghost father. I thought, *Could this be my avenue?*

My personal search for my origins had grown old, stale, and dormant since my move to Philadelphia. I felt like I had already turned over every stone to no avail. I had learned what I could about the history of artificial insemination by donor and the social context surrounding artificial insemination by donor during the time of my birth. I had followed all the clues provided by my mother and her stroke-impacted memory; Dr. Sims, the doctor's fertility practice in Boston at 10 Beacon Street, and Harvard Medical School, all had led to nowhere specific. And my mother was no longer alive to provide a sounding board, even one with cognitive limitations, for any new discovery I might have made. I felt very alone in a wasteland of nothingness.

Until DNA testing, I had no other place to turn. No sperm bank to survey. Nothing!

If 23andme could uncover, at the very least, my paternal genetic heritage and unlock any health mysteries specific to my DNA, signing up to it would be well worth the price of entry. Finding the "who" of my paternal seed was beyond my expectation. I was looking for the "what."

Let's give it a shot, I thought.

In early 2008, just after my sixty-second birthday, and with nothing to lose except the $999 price tag, I excitedly spit into a specialty 23andme vile to become one of its early group of customers, proudly part of the "Lunatic Fringe."

During that "as advertised" eight-week wait for the results, the game plan to turn my sick company into a healthier version of itself struck early success. Its stock value skyrocketed. With new cash in the bank, my firm was on its way back to health and prosperity. I was also an emotionally healthier version of myself; uncovering the secret of my origin and successful therapy had given me the tools I needed to carry my baggage and enhance my working relationships.

I had changed fields, practiced some leadership fundamentals, took a few bold moves, assembled a terrific team of people, and pulled off a major turnabout of fortune. Celebration was in the air in my professional life.

Or so I thought. The financial crisis that began in late 2008 was preparing to disrupt those early achievements in a nanosecond.

The success that had been achieved with the company early on was temporarily rewarding, but I was distracted. After the first month of waiting for my test results, I checked my personal email account for an email from 23andme every day, sometimes several times a day.

My old self might have silently felt anxious and avoided discussing the confusing emotions. Am I the product of a Jewish doctor from Harvard Medical School? I understood that I was maternally part Jewish, Scottish, English, and French. What about Northern Italian? Would 23andme magically uncover a close relative who could identify the donor of the sperm used in my conception? What might be the positive and negative inheritance factors that I would learn from this genetic analysis?

My new self shared the state of my psyche with my family and close friends, who listened without judging me. That both soothed me and fortified my patience.

Finally, near the end of the seventh week, I sat at home behind my laptop with a freshly brewed cup of Peet's mid-roast coffee at my side. At 6:15 a.m. on a flower-scented Friday morning in spring, a hint of sunshine appeared through my den window, and then from my email.

There it was, finally. "Your DNA results are available. Click here and log in."

Chapter 17

Such emotional turbulence. I had so many questions to be answered and no idea what, if anything, I would learn. What was my paternal heritage? Were there any health issues in my DNA? Did I have any close relatives who could help identify the ghost father who contributed his life-giving seed to enable my conception?

There it was at last, an unfamiliar web page to figure out how to navigate. What to click first: Ancestry, Health & Traits, DNA Relatives? The blue oval begged for a click. "View your ancestry composition."

Click!

A new screen popped up. "Your DNA tells the story of who you are and how you're connected to populations around the world."

No kidding! My dreams of instant gratification were immediately dampened, and what little patience I possessed was taxed by dealing with an unfamiliar website. I had been a CEO for too long, I thought; too many people did these tasks for me. I had not honed the skill set on my own. I clicked on "Ancestry."

The summary of my total ethnic heritage was very similar to my understanding of my mother's heritage: British and western European (most likely French) from her father, coupled with Scottish and a piece of Ashkenazi Jewish from her mother.

But one addition stood out: not Northern Italian, but Scandinavian.

The site also highlighted that my DNA contained a far smaller Neanderthal component than 86 percent of the 23andme customer base.

"That's comforting," I sighed. "I'm not a Neanderthal after all," I told myself. I later joked to my inner circle, "How would that look on my resume?"

The site's icon allowed me to drill down further on either my maternal or paternal "haplogroup." I do not have a degree in molecular biology. I had no idea what this term meant. Google to the rescue! Feeling that I was, in fact, more Neanderthal than 23andme had determined, I hastily typed "What is a haplogroup" into a secondary browser so that I could have two websites up at once. The response: "A haplogroup is a group of similar genes inherited from a common ancestor."

With a click on the paternal icon, 23andme provided a scientific narrative describing the migration of my genotype over the past 275,000 years, accompanied by a color-coded map of the globe to show that migration.

I took a sip of now-lukewarm coffee and said, "Oh, for Chrissake," rather loudly to both my coffee cup and my aging Dell. It took a bit more dissecting of the scientific language to gather that the I-M253 haplogroup was rare among 23andme customers, 1 in 250,000. Its Scandinavian origins are commonly found in 15 percent of men, primarily those from Sweden and Denmark and secondarily those from England, Germany, and the Netherlands. Norman invaders had apparently fathered many children throughout their two-century-long European rout, I reflected, as I recalled my world history studies.

Next, the site offered a rather detailed explanation of the Y chromosome that determined my gender. I was satisfied that I was a male.

I neither wanted nor needed a refresher in high school biology at that particular moment.

Interpolating my knowledge of my mother's ethnic heritage with 23andme's summary of mine, it seemed reasonable to conclude that my ghost father was, like my mother, a composite of British Isles (English, Scottish) and western European (French or German). But rather than being a sliver Ashkenazi Jewish like my mother, my biological father had been a sliver Scandinavian.

According to the provided nicely color-coded bar charts of all twenty-three pairs of chromosomes, I also deduced that approximately two-thirds of my inherited traits had derived from this Scandinavian ghost and one-third from my mother.

My second cup of coffee was cold by the time I clicked the icon that read "Trace your paternal line: Visit DNA relatives that might be specific to your paternal line."

My heart thumped as though I had just finished a five-mile uphill run. Would I find a real person who could help identify my ghost father? A big NO.

Every single one of my relatives in the 23andme database appeared regardless of whether I clicked "Trace your paternal line" or "Trace your maternal line." It bunched all of them together; they were not segregated by father or mother as the service had claimed they would be.

At first blush, I was DNA-connected to several hundred very distant cousins; just two were second cousins. From an initial review, their connections appeared to trace to my Canadian maternal grandfather. He was one of fifteen children. His mother and father, as well as his grandparents and their parents, all had a dozen or more siblings each. "Rabbits," I joked out loud. He was the only member of his extended Canadian family who had immigrated to the United States (Boston) from Nova Scotia, which he did as a young adult in 1910.

A calendar reminder intruded on my 13-inch Dell screen at 7:45 a.m.: "Happy birthday, Kitty!" It was my mother-in-law's eighty-eighth birthday. A loud "Chrissake" surged in a growl from the back of my throat. I had run out of time.

A packed Friday calendar was always common after I returned from a few days of mid-week business travel. I needed to get to my office for a series of scheduled meetings with would-be investors and a presentation from a company in which my company was contemplating taking an ownership stake. Finding whatever answers 23andme could divulge regarding my Health & Traits would have to wait.

"How about a hot cup of coffee?" asked the cutest, most endearing woman I was fortunate enough to have in my life. Susan did not have to ask about what I had just learned. She handed me a fresh cup, pulled up a nearby hassock in the den, sat beside me, and held my hand.

"Looks like the ghost was mostly British, some western European (French or German), and, how about this, Scandinavian," I offered. "That explains my dad's Northern Italian blue eyes, along with someone else's fair complexion, lighter hair, and easy sunburn. No paternal relatives at first pass."

Susan smiled in her special way, tenderly stroked my arm, and said, "Good for you! I love you for who you are. Now you know the 'what.'" As she added a kiss for good measure, she said, "You've come this far in your search. There's no telling how much farther you can go. Or this could be it. I hope you're okay with that."

I replied, "Yes, this might be all that I learn . . . but then again . . ."

Would it be enough for me? At that immediate moment, I just did not know.

While driving to my office, I made a flurry of cell phone calls to my inner circle. "Hey, you'll never guess what I just learned."

Humor, I had discovered throughout my difficult times, especially in combat, had diffused the anxiety of being in a tough spot. My old high school mentor and former social studies teacher, Owen, started calling me "Sven."

I told Mike, who was a 6'2" Swede, that he was dead wrong. "My ghost father wasn't a Jewish doctor; he might very well have been a Swede, too." With a laugh, I added, "And I received confirmation that I'm not a Neanderthal."

Logical cousin Eddie, the pharmacist, said, "This is the coolest thing imaginable. Who knew?"

Throughout the Friday evening and Saturday morning of what I labeled "the 23andme weekend," I immersed myself in the 23andme site and clicked on every icon that they provided, sometimes several times, to better learn to navigate the site.

I focused first on "Health & Traits," and then moved to "DNA Relatives." According to the DNA health analysis, I carried no glaring susceptibility to specific diseases—no likely inherited heart disease, cancers, diabetes, or the like. I breathed a sigh of relief. I could inform Tracy and Stephen that I was not a carrier of something awful they might have inherited.

I exhaled my second breath as a sigh of anxiety. Where did mental illness fall in my inheritance cycle? That did not show up, but neither I nor my adult children had shown any signs of the debilitating unipolar depression that had plagued my dad.

My lifelong athleticism and muscular conditioning also pointed to a positive influence from my DNA. The report stated: "Your genetic muscle composition is common in elite power athletes." That icon drilled down to a more scientific explanation if I wanted it. Reading more slowly now, I devoured all the scientific information and wondered silently whether my parents' fertility doctor was a eugenics disciple. There was no guarantee that my inherited health prospects were really that clean, but I was pleased with what I had found, or, said another way, also pleased with what I had not found. So, too, were my adult children.

My Philadelphia primary care physician, Jon, and I had earlier discussed my pending 23andme test and what the results might show. Jon said, "DNA information is credible, but outliers always exist. These tests capture just a piece of the gene variants that can lead to disease."

I sent him my DNA Health & Traits summary early the following week. He emailed back to say that the information I had gathered did not raise any red flags. He closed with, "Continue to practice preventive healthcare maintenance."

The limitations to what I could glean from the site emerged like a jagged boulder that appeared at low tide on the approach to an unfamiliar harbor. It altered my course and shadowed my wonder and enthusiasm inside of that first weekend. Then I saw that 23andme encouraged me to enter relatives' details, names, and birth dates to help determine genetic relationships with other customers in the database.

What names? I had no paternal names. Then I remembered that maybe I did.

Tucked away in the back of my briefcase was that bright yellow folder I had uncovered in my research with my daughter that had the names of 1945–1948 Harvard Medical School graduates. Like a man possessed, in a frenzy, I dumped my briefcase upside down, emptied its contents onto my den floor, scooped up the yellow folder, and monopolized our dining room table for an hour.

I checked off the names that appeared the least bit English, French, German, or Scandinavian and then did a Google check on their origins just to avoid any glaring errors. That, at least, winnowed the potential field of sperm donors down by about half.

Upon entering a couple of names just to test what might happen, I likened the experience to looking up the name "Smith" in a city telephone directory—finding a sea of names with no clue as to which Smith to call. If I had called everyone named Smith, what would that tell me, other than their name was Smith? If there were a Smith to which I had a genetic connection, how could I tell if it was paternal? With no real paternal name to enter on the 23andme site to begin with, I felt dead in the water. This avenue to discover more about my ghost father remained tightly shut and locked. Disappointingly, a paternal DNA connection that could provide a clue was still missing.

Susan, in the meantime, paced impatiently waiting for me to finish so that we could drive seventy-five miles and arrive on time at a Lehigh Valley restaurant for her mother's birthday luncheon with a small gathering of her extended family.

While smiling and affable throughout the family celebration, I felt preoccupied with both this new discovery and the absence of any breakthrough on the identity of my ghost father. I escaped to a corner of the party room with my brother-in-law, Bob.

Susan's family was also my family. We enjoyed each other's company. I had become a close uncle to Bob's three children, with whom I had bonded during their Cape Cod visits throughout the years. Bob was part of my inner circle. I updated him on my new discovery and the new wall preventing additional genetic disclosure.

My friends and family were always bright and enthusiastic over any new detail I had uncovered. Bob buzzed with excitement for me, but I felt alone with what I did not know. Who could identify with the emotional void, the bewilderment, and the emptiness that I felt deep inside? Perhaps an adoptee, but I was not close enough to anyone with that background. Susan was always perceptive. It wasn't enough for me to know the "what." My insides demanded to know "who."

Loneliness was not new to me. Leadership often occupies a lonely seat. As a CEO leading troubled entities out of hot water, I was accustomed to feeling lonely on occasion when facing complex situations, competing interests, and the need to make tough calls. I had developed a support system to cope with those stresses. I enjoyed a CEO circle of friends and acquaintances who could relate; we could commiserate, and I found that camaraderie both supportive and helpful.

But I kept the donor-conception piece of my story under wraps; we discussed business. I knew that a handful of donor-conceived others like me were out there, but I expected they would be a rare find. Of the 0.00014 percent of the population that was donor-conceived, I thought, so few even know that they are. I longed for a conversation with someone who had walked in my shoes, who had experienced these emotions, someone who could relate. In 2008, I could find no such circle.

My knowledge of my donor-conception magnified my loneliness immensely, as did the never-ending nature of my search, the breakthroughs of discovery, and the subsequent impasses; I had no donor-conceived camaraderie to lean on. I felt stranded on a remote island, alone, longing for a colleague to share the experience with.

Professional drama and unprecedented complexity in my workplace were about to become a welcome, consuming distraction.

As 23andme unveiled additional details of my genetic identity, a rumble of economic noise in Europe had traveled at supersonic speed. A volcanic eruption was to be felt the world over. A proliferation of bad loans abroad had created a European banking crisis. But it was given little note in the United States as the Dow Jones Industrial Average index had hit a then-record 14,000 points in October 2007.

But, over the span of several months, a meltdown in the sub-prime mortgage lending market compounded global banking fears. Several US banks collapsed. Home prices dropped overall by 20 percent, even more in some harder hit metropolitan areas.

By the autumn of 2008, Lehman Brothers filed for the largest bankruptcy in US history. The stock market was in free fall and dropped 500 or more points per day for days on end. Americans lost over 25 percent of their collective net worth as the stock index bottomed out to just below 6,500 points, a 54 percent decline. "Credit crunch" and "bailout" became household terms that threatened the survival of auto makers, appliance manufacturers, small businesses, and individuals.

In October 2008, as 23andme offered the identification of "predispositions to ninety different traits ranging from baldness to blindness," the company gained recognition as *Time* magazine's "Invention of the Year." I barely noticed amid all the chaos. I had other things on my mind.

Unemployment soared and eventually exceeded 10 percent—twice the pre-crisis rate. The directors on my board were shaking in their boots; the company's stock had again plummeted to under a dollar per share. Main Street, Wall Street, and Pennsylvania Avenue were all consumed with how to best deal with this imminent collapse. The "Great Recession" threatened everyone. Who had time for 23andme?

For a few tense years, my "leadership through adversity" skills were put to a huge test. I used my bluff bravado to camouflage my own fear as I repeatedly encouraged my colleagues: "The measure of who we are isn't how high we are when we're up, but rather how high we can bounce when we're down." We executed our business game plan with the required flexibility, focus, and intensity. With collective wisdom, bold moves, and tough negotiations, coupled with a few setbacks along the way, my team not only navigated through this death trap, but earned top-tier VC returns, paid off our choking debt, and stashed record-breaking accumulated cash in corporate coffers. Over a two-year period, from the depths of despair until the stock market bottomed out and began its march to recovery, the business's health and stock price vaulted to a post-bubble high.

The camaraderie, celebration, and feelings of accomplishment were not only richly rewarding, but they also provided a perfect platform from which I could make an important life change. I announced to my family that I was forming a retirement plan that would take a year or more to complete. Board preparation and succession planning were pieces of it, as well as the maturation date of my stock options. I also had to plan for my own needs. At age sixty-seven, I knew that it was not emotionally healthy for me to grind to a standstill after more than fifty years of cycling at such a rapid pace. I could not just call it a day.

Reality therapy? I thought. Now I could reflect on my feelings and at the same time come to rational decisions without denying those feelings. The grind of 100,000 miles a year of travel, regular breakfast meetings, and board room and shareholder dynamics had all intruded on my enjoyment of "the good life." I needed a transition plan.

I retired in 2013 at the top of my game with plans to use my time to write a book (*All Hands on Deck*, published in 2015), start a consultancy (Kedgeway LLC), chair a global nonprofit (the Network for Teaching Entrepreneurship), sail my newly acquired 31-foot sloop (*BluBird*), and spoil my growing family. By this point, our kids were both married to wonderful partners and had provided Susan and me with a few beautiful grandchildren. I was ready to turn my page to a new chapter, become someone I had never known—a grandfather—and continue to search for the unknown details of my conception.

Over the next few years, I logged in to 23andme repeatedly with fresh hope and anticipation that I would discover new information or find close paternal genetic connections to new subscribers. Over the years, 23andme had proactively emailed with updates to my genetic profile or notified me of new DNA matches as their customer base grew.

Using the site's email messaging capability, I had connected with just one "distant relative" who also carried the I-M253 haplogroup: a PhD in microbiology. I gave him a brief sketch and asked if he could provide a clue to the source of my ghost father. He emailed me quickly, with the depth of knowledge that one might expect from a microbiology PhD. He pointed out that haplogroup had little to do with genetic relationship. I joked that "perhaps we shared Norman invaders in our DNA." Just as quickly, by looking at names on our respective trees, we determined our connection was on my mother's Canadian side. The prospect of finding a paternal connection felt all the more frustratingly unachievable to me.

Others messaged me and offered a deeper history of my maternal, Canadian roots. Interesting, but not what I sought. Over the next few months, 23andme continued to evolve and improve its look, feel, and content, but each visit proved repeatedly empty of paternal discovery. Susan's comment at the beginning of my 23andme discovery reverberated loudly within me: "Or this could be it. I hope you're okay with that."

I was not.

Chapter 18

A few years would pass before anything new happened on the genealogy front. Consulting projects, the launch of my book, promotional speaking engagements, leisure travel, recreational seasonal living on Cape Cod, and regular family visits to New York City and San Francisco to see our grandchildren filled our time and wonderfully captured our attention and interest. I was living the good life. I was active, healthy, financially independent, happily married, and enriched by a beautiful family . . . but all the same, I had an unfilled genealogical void. I felt guilty that even though I was so fortunate, I also felt unsatisfied. Who could understand that?

On Thanksgiving weekend, 2016, our Philadelphia living room was ablaze with festive bedlam. Tracy's youngest, my one-year-old grandson, was crying over being annoyed by his older brother. That three-year-old brother and his four-year-old cousin were giggling hysterically as they played Whack-a-Mole with a wooden mallet.

Ah, family!

Amid the commotion, I scrolled through channel after channel on the television to tune in to a couple of college football games. Only cryptic adult conversations could penetrate the chaos of the pending grandchildren's bedtime.

"Anything new with 23andme?" Tracy asked.

I did not give voice to the burning emptiness I felt or the guilt I carried about living a perfect life while still feeling hugely under-accomplished and dissatisfied when I answered her: "Perhaps I've reached the end of my search," I answered. I felt so discouraged. "I want more. I've learned so much, but it just might be the end of the road. I guess I can live with that."

"Yeah, right," said the young woman who knew her father all too well. "Let's find a different road."

Instead of giving us a college football update, the television barraged us with an unrelenting dose of Black Friday advertising. Joining the fray, Ancestry.com ads saturated every channel. They offered Black Friday discounts off their $99 list price: on sale for $49.

It occurred to me that DNA science, like other scientific advances, had gathered Category 5 hurricane force over the warming waters of artificial intelligence (also dubbed AI). Its computational power doubled every three and a half months, soaring seven-fold past Moore's Law (named after Gordon Moore, the CEO and co-founder of semi-conductor giant Intel), which predicted that chip density would double every two years. Chip capacity had an enormous impact on the price and performance of technology products like mainframe computers. Once priced at $1 million, the power of a mainframe could now fit in the palm of a hand for under $1,000. Moore's Law had seeped into DNA tests, which were now at a price some 90 percent less than they had been during the previous decade.

"Just eight years ago, 23andme set me back about $1,000," I observed. "Now, this stuff is selling for less than fifty bucks."

In between the background commotion of cries and squeals, my son, Stephen, explained about another DNA analysis option: "Ancestry.com used to just offer birth, baptism, marriage, and death records. Stuff to construct a family tree. But they're now into DNA analysis in a big way."

Have I become so old and stale that I haven't kept up with these advances in DNA science? I asked myself.

Stephen gave me a further education, explaining that over its first quarter century, Ancestry.com did business under varying trade names, had merged with or acquired complementary businesses, and operated under several ownerships. First owned by various publishers, it became a publicly traded company for a time, and then was taken private once again by VC/private interests. Initially offered by Ancestry.com in 2012, DNA testing appeared to be a terrific and timely new product to sell as an add-on to its current subscribers. Its customer DNA database was surpassing that of industry pioneer 23andme, which had a five-year head start. Ancestry.com was going to prove instrumental in my discovery of where I came from, even though it would be several months after our Thanksgiving discussion that I would take the test myself.

Adult conversations can stall; bath time, grandparents reading bedtime stories, and getting three excited young children to sleep are all-consuming. It took several months for us to get back to the topic.

We moved into our Cape Cod home for the 2017 "season" in late April. The boatyard notified me that they were running a few weeks behind schedule to launch *BluBird* due to an unusually wet spring. By mid-May, I had enjoyed several gatherings with old friends from high school and college who lived in the general vicinity—there are no friends like old friends.

Then, quite unexpectedly, Tracy had a breakthrough that established a beachhead for an emotional four-month ride of new discovery that led me to the promised land.

Sent: Friday, May 12, 2017, 9:01 p.m.

Subject: John Rock

Dad, check out this link from the NY Times: *John Rock, Developer of the Pill and Authority on Fertility, Dies*

Tracy had sent a *New York Times* obituary, dated December 5, 1984. I clicked on the link. The obituary stated that Dr. John Rock had died in New Hampshire the previous day from a heart attack. The author noted Dr. Rock's pioneering work on human fertility and the development of the birth control pill. Of particular interest to me was the description of his work on human eggs and sperm: "Dr. Rock was the first scientist to fertilize a human egg in a test tube, in 1944, and among the first to freeze sperm cells for a year without impairing their potency."[9]

I marveled at how she was able to keep coming up with these new findings. She is married, has two young boys, is professionally employed, and yet continues to feed me enlightening information to fill my void. I realized that I had spent more than enough of my post-retirement time consulting, promoting my book, and chairing a nonprofit. All those activities helped me distract myself from facing this void—and avoid facing the fear that I had traveled to the end of my road of discoveries.

Before the weekend ended, Tracy had sent me another gem. *She is Wonder Woman,* I thought. *She is an expert sleuth always uncovering something interesting and providing an internet link for my easy access.*

---

9   Joseph Berger, "John Rock, Developer of the Pill and Authority on Fertility, Dies," *New York Times*, December 5, 1984, Section A, Page 29.

Sent: Sunday, May 14, 2017, 7:02 a.m.

Subject: Check this out on Amazon.com

Dad, I think this Rock guy had a clinic or was affiliated with one in Brookline that did artificial insemination in the 40s. Here's a link.

The Fertility Doctor: John Rock and the Reproductive Revolution (John Hopkins University Press)

This one was a 2008 biography of that fertility doctor. With a click, up came the Amazon.com summary, which offered information similar to what had been listed in his obituary. I was rather dismissive in my reply. In vitro fertilization (IVF) was quite a bit more sophisticated than my conception via a mere sperm donor.

When my girl gets a hold of something, she has never been one to be denied. She pushed her point, quickly writing back, "Yes, but he was working in fertility before he attempted IVF. It's more about who his colleagues were and what services his clinic offered."

As a rainy morning slipped to a rainy afternoon, she continued to send me links. With no boat in the water, no company in the house, heavy rain, and not even a Red Sox game to watch, this seemed like a productive use of time. I clicked on one of the links she sent: *John C. Rock Personal and Professional Papers: 1921–1985 Inclusive*. It read: "The John Rock Papers, 1921–1985, are the product of Rock's activities as a fertility specialist and endocrine researcher at the Free Hospital for Women in Brookline, Mass." What a trove of information! Most importantly (to me, at least), the collection included patient information and a variety of materials related to his work on artificial insemination. But as I read on, I hit two roadblocks. First, patient names had been redacted to protect confidentiality. The second problem was this: "Access to personal and patient information is restricted to 80 years from the date of creation."[10]

Eighty years! That was nine years away. Pray that I even live that long! On top of that, my mother said the clinic's address was in Boston, not Brookline. I googled to get more information about the hospital. It had been founded in 1875 by Dr. William Henry Baker, who had been a resident under Dr. Sims at the Women's Hospital in New York City.

While Tracy's two young boys napped, she and I spoke on the phone to discuss her findings and mine. This was our very first link to the mysteriously missing and now long-dead Dr. Sims. I reminded her of the secrecy of the times that I had uncovered in my earlier research. What records might exist? And Dr. Sims's artificial insemination research that happened in 1866 might be historically interesting, but I was conceived in 1945. Culture at the time encouraged so much secrecy around the procedure that the doctors may not have kept clear records of the patients, much less the donors.

Tracy used her research findings to circle back to our Thanksgiving conversation in Philadelphia five months earlier. "Dad, when are you going to sign up to Ancestry.com? They have really grown their database. Now, 23andme isn't the only game in town. Perhaps you'll hit a connection on another fresh database of DNA customers."

Rainy, sleepy Sundays generally offer a fine time to catch up with adult children who live in other time zones. My son called from San Francisco late in the afternoon the very next day to do just that. Of course, as I later learned, the siblings had colluded.

Stephen reinforced his sister's Ancestry.com message. "A list price of $99 is a mere fraction of the $999 you first paid." My hesitation was not about the money.

---

10   John C. Rock personal and professional papers, 1921–1985, Francis A. Countway Library of Medicine, Center for the History of Medicine, Harvard Library, Boston.

He repeated what we had already acknowledged. "Since you discovered a good deal of the 'what' via 23andme, perhaps you can find the 'who' after all. Advancing science has surely altered the pre– and post–World War II guarantee of sperm donor anonymity." He concluded, "Ancestry is a much better avenue for you right now, Dad. Standing still isn't your speed. You'll never feel settled with this unless you continue to advance."

One of his words evoked the key premise, "advance," that I had used in my book, *All Hands on Deck*. I was solidly stuck in the mud, making zero progress, even though I had built my personal and professional persona on being the right guy to get things unstuck.

My kids clearly understood how to push their father's buttons. They were right. I was afraid of more disappointment if I continued on this journey.

Over the previous few years, a continuous stream of DNA stories had hit the media. At first, they were uplifting memoirs of adoptees finding their roots and meeting blood relatives. Those stories were followed by sperm bank donor–conceived stories that made donor anonymity obsolete and identified dozens of half-siblings conceived from the same donor sperm. Reports of those reunions were not uniformly heartwarming; often they were stories of rejection. In one case, a 1949 pre-sperm-bank-conceived man (Bill C., from the 1995 *New York Times* story that Liz had copied for me) discovered over sixty half-siblings, all originating from the sperm of the same fertility doctor, all within the same community. Reproductive science had continued its advance. Egg donor–conceived and frozen embryo donor–conceived individuals joined the fray. Each article contained similar themes: advances in DNA science that made the standard practice of anonymity obsolete and an unabashed absence of reproductive industry regulatory authority, which enabled fraud and abuse.

Yes, I was stuck; I was afraid of what I might discover. But discover it I must.

On May 16, the ninth anniversary of receiving my 23andme DNA analysis, some twenty-two years after my mother's secret first came to light, and twelve years after her death, and for all of $99, I ordered a second DNA analysis, this one from Ancestry.com. (Speaking of good DNA, on that very day my mother-in-law celebrated her ninety-seventh birthday in a private room in her Pennsylvania nursing home with her three siblings and two children present to light the candles on her favorite vanilla cake.)

The spring rain had finally subsided. That summer, the beach house's crushed-shell driveway received a steady stream of friends and family who piled out of their vehicles to share the joys of retirement with us.

Central to our lives were the late spring and summer months of Cape Cod sailing, beaching, swimming, exploring, eating, drinking, and partying; this retirement thing suited us well. We transferred the holiday season's bedlam from our Philadelphia living room to the entire beach house for the summer season.

In between visits from family and friends, our pattern was to take a few days reprieve to sail without guests, give the house a major scrub, and stock up once again for another round of early August company. My children and their families would soon be crunching their tires on the driveway shells with their own honks, waves, and anticipation of fun in the sun.

Mother Nature abruptly intruded upon our summer festivities as storm clouds coalesced with a turbulently churned rogue wave as large as any created by a Category 5 hurricane. The 2017 Atlantic hurricane season captured its indelible place in history; it had seventeen tropical depressions (ten of them at hurricane strength) and the fifth-most named storms since 1851, when reliable record-keeping began. Among those storms were six of the most savage storms, Categories 3, 4, and 5, all of which hit landfall in rapid succession. Cape Cod received no direct hits, but it experienced its share of the remnants of those storms in the form of humidity, fog, rain, some flooding, and gale-force winds.

Near the end of August, Hurricane Harvey plowed into the Texas gulf before traveling northeast as a tropical disturbance. In September, Hurricanes Irma and Maria devastated the Caribbean before making damaging landfall in the Southeast. Hurricane Jose traveled up the eastern seacoast, touched Cape Cod with its outer circle, and downgraded to a tropical storm as its remnants punched Woods Hole. By early October, Hurricane Nate perpetrated the worst natural disaster in Costa Rican history. The World Meteorological Organization retired the names of those storms due to their record number of deaths (3,364) and catastrophic amount of damage ($3 billion).

Yet, hurricane history does not record a memorable name given in the midst of these record-breaking Atlantic whirlwinds: Morgan.

On a rain-drenched Sunday morning in early August, I settled in our casual and comfortable Cape Cod living room in the pale blue upholstered swivel rocker next to a fieldstone fireplace. Hidden from view under the rocker's left arm was an antique cranberry box where my laptop often rests. With my shoeless feet elevated on the chair's matching hassock and a cup of my old friend, Peet's Café Domingo, in my right hand, I reached for the laptop to get caught up on my email. That was the beginning of an emotionally wringing two months of emails, texts, phone calls, meetings, and new discoveries.

I had barely noticed the Ancestry.com email sent to me in mid-July. I had read it rather hastily two weeks earlier:

"Thank you for registering with Ancestry. Your tree and Ancestry's family records help you to discover your unique story. With your AncestryDNA test, you will discover more about your ethnicity and your living relatives."

Would Ancestry.com come to the same ethnicity conclusion as 23andme? How about identifying a paternal relative? The blue rectangle urged: "Explore AncestryDNA."

I clicked on "Ethnicity and Origins."

A color-coded map of the globe appeared; next to the globe was a column that detailed Ancestry's view of my ethnicity and a multi-colored pie chart.

While the numbers were not exactly what 23andme had reported, it was close enough to 23andme's data to corroborate my ethnicity and the ethnic composition of my ghost father: English, French, and a slice of Scandinavian. These results specified "French" versus "western European." I became more aware that as DNA databases had grown, the DNA analysis firms had more data to work with and were better able to pinpoint ethnic origins.

Unlike the 23andme history of human migration since the Ice Age, Ancestry.com narrated their rendition of immigration history to North America, which I hastily scanned:

- English, Scottish, and French settlers to eastern Canada and the historical backdrop during that time—*Check!*

- English and western European settlers to Boston, along with the history of the times—*Check!*

What came next so disgruntled me—erroneously so—that I logged off the Ancestry.com website and wrote it off as an inadequate experience.

Ancestry.com had prominently promoted their family tree tool throughout the site, and each click attempted to upsell me. For another $99 over six months, I would gain access to Ancestry's database, which would include newspaper clips, obituaries, census records, draft board lists, baptisms, birth certificates, death certificates, and marriage certificates. All that was needed was a real name, not a "Smith" in a sea of Smiths. What name? Without a real name and a date to accompany it, the family tree feature was absolutely void of paternal-discovery value for me. Such a disappointment!

The Cape Cod weather forecast called for a window of weather at its best, and my grandchildren were about to arrive. I logged off the site, put the aggravation behind me, and spent the next few weeks enjoying family, beaching, mingling with friends, and sailing.

Tracy approached me over the course of her visit with the wisdom of one who knows her father's impatience with unfamiliar websites. "Once the summer activities settle down, spend some more time on the Ancestry.com site and see what you find." Stephen also encouraged me to pick up the search again.

Although she was not encouraging, Susan remained tolerant and supportive of my continued research, but she was concerned that I had, indeed, traveled to the end of the road. Had that day come? Would I be okay with it? Apparently not!

The waning days of summer on Cape Cod take on a peaceful, serene charm. Vacationers generally clear out as Labor Day approaches. Our beach house was empty and quiet for the first time since the end of June.

The three-week siege of spectacular weather was finally interrupted by a humid, windless afternoon. Rain and winds from the remnants of distant Hurricane Harvey were in the forecast. It was now a perfect time to log in to Ancestry.com and repair my earlier negative experience with a more positive and patient approach.

I finally recognized my incorrect interpretation of what Ancestry .com had to offer. A whole new level of discovery then exploded like a howitzer artillery shell; it shook the earth under my feet. With a few clicks, I noticed "View DNA Matches."

How did I miss that the first time? I clicked on the link.

There she was.

Metaphorical sirens blared. Strobe lights flashed. I sat, calm and still, on that soft, comfortable living room swivel rocking chair with my Mac in my lap. In a sea of distant cousins, she was the first listed:

"Roxanne, Close Family, 1st Cousin?"

First Cousin with a question mark. What was that about?

My mind raced with possibilities. I grew up visiting all my maternal cousins, the children of my mother's brother and sister. Given my knowledge of my maternal family, Roxanne had to be a paternal cousin.

Was this the pay dirt score about which I had been fantasizing? Would she embrace me or push me away? Could I be perceived as a bastard stepchild seeking a seat in a king's court?

There was the messaging icon. Just like 23andme, one can send a private email to a connection. I clicked on it, and then I froze.

What should I title this? I did not want to give too much information all at once. I worried it might frighten her off. Then I stopped worrying. I had come this far. It was time to go all in.

Sent: Wednesday, August 29, 2017, 5:17 p.m.
Subject: SEEMS THAT WE'RE 1st COUSINS!

Hello, I joined Ancestry to discover more about my paternal roots. Quite a mystery so far! British, French & Scandinavian heritage w/ a connection to Harvard Med School and Boston in the mid-1940s. My maternal roots combine Eastern Canada, Scotland, England, France, with a piece Ashkenazi Jewish. Given my knowledge of my family, our connection appears paternal, but who knows? Can you help fill in some blanks? Thx.

I spent the next three weeks glued to that comfortable living room swivel rocker in the semi-permanent position of "Mac-in-Lap." With the exception of the occasional day sailing, *BluBird* remained hunkered down on its sheltered-harbor mooring. Hurricanes had hit the pause button on Cape Cod sailing and the pause continued through the close of the season. With that circumstantial backdrop, my new-cousin discovery was just beginning. None of my mariner, military, or business training prepared me for navigating through the personal, emotional roller coaster ride of my life that lay ahead.

Chapter 19

After cleaning up from an early dinner and watching NBC's Lester Holt on the evening news, even though only two hours had passed since sending my first message to this new cousin, I grabbed my MacBook Pro to check my messages. When I opened it, I was delighted to find that—despite all the time it had taken to get me to this point, literally decades since I first learned my parents' secret—it took no time for "Roxanne, Close Family, 1st Cousin?" to respond to my email. I had been emotionally preparing myself for rejection. Was I to be perceived as a threat, ostracized, and ignored? She surprised me with an immediate warm and welcoming embrace. Her reply was friendly and pleasant, and she seemed eager to know more about me.

I initially learned that Roxanne's eldest son gave his mother the DNA kit as a 2012 Christmas present. She was among Ancestry.com's first customers to view her DNA analysis. Using its "open access" feature, she welcomed other Ancestry.com subscribers to view her deeply detailed family tree. She admitted to being an Ancestry "aficionado," with a congenial and helpful greeting to all would-be distant relatives. She had no idea how close a distant relative was about to become.

Roxanne told me she had recently picked up the mantle as her old-Yankee family's historian. In 2002, she began digitizing the results of seventy-five years of her mother's research into her family tree and had handily mastered the use of Ancestry's online records locator to dig even deeper into her family's past. She had details of her family's old New England Yankee history. She told me, "All of my ancestors have long New England roots, with some back to the *Mayflower*."

I would come to discover that embedded in her proud lineage were such dignitaries as her namesakes: her eighth great-grandfather, Roger Williams, the Puritan-dissident founder of the Colony of Rhode Island; Sir Francis Drake, a paternal cousin several generations removed; and several dukes, earls, knights, and whatnot from royalty in old England and western Europe. Talk about a pedigree. For so many years, I had struggled to prove my flawed self equal to peers who had vast pedigrees to boast about. Now there was a real possibility that I shared quite an impressive genealogy in a family I had no idea existed until Roxanne replied.

She explained, "I have done extensive work on my family tree and would love to see where you are in my tree. Your last name has me stumped since I have no such surnames in my tree. Also, are you a New Englander? I am very curious about your Harvard and Boston connection also," she wrote.

It seemed doubtful at the time that she might be able to help me because she asked me for information I knew was not going to be helpful—my parents' names and birth/death dates, grandparents' names and birth/death dates, and where they lived. Even knowing my age was not going to help too much in this mysterious origin story of mine. I knew I needed to come completely clean and tell my unlikely sperm donor tale, and do it in a condensed but compelling form. I began:

I'll share a mystery with you that I learned from my mother when she was 75 and I was almost 50. I've also shared this mystery with my immediate family. That was 21+ years ago. I'm 71 now. The man that I knew to be my father wasn't my biological father.

Inside of two paragraphs, I unloaded a brief version of my mother's secret, along with my 23andme experience. In addition, I outlined my early biographical history to this new cousin and included all my contact information, links to my consulting practice's website, my LinkedIn profile, and the title of my first book with a link to my Amazon author's page. The cyber world is a hostile place with scam artists galore. I felt it was important to state that I was not looking for anything from her, just information. *If I am totally transparent*, I thought, *I will appear more credible.* Susan served as my copy editor before I hit "send."

Roxanne replied within the hour.

What a family story! I have two relatives who were physicians in New England. I will have to do more research on them to see if there was a Harvard connection. Both my parents were only children. This must mean our DNA connection would go to siblings of my grandparents, is that correct? What an unbelievable connection that we have!!! Hey, you have some really good DNA :)!

"The ghost is a doctor. He's in her tree," I blurted out immediately. "Holy shi—!"

My cry echoed across the salt pond to a flock of passing Canadian geese and an assortment of feeding egrets. I could taste the victory of finding long-sought knowledge. It was within my grasp. "Fight to the last gasp," I said to myself. "Never doubt my resolve."

Forty-three minutes later, Roxanne wrote back: "Could my great-uncle be a match in this search? I have been working on this diligently and will continue tomorrow."

I have the gift of being able to sleep through anything: a mortar attack while in the jungles of Vietnam or a sticky business situation under the stress of impatient and demanding public shareholder scrutiny. But not that night!

My mind raced with possibilities. Was his name on that list of Harvard Medical School students that I had stashed away in my briefcase? I scrolled through the names I had memorized as I lay awake, excited and anxious about what I would learn in the days and weeks ahead. John A. Williams stood out, a graduate in 1946.

That morning, it appeared that the remnants of Hurricane Harvey had all blown by. A blue sky emerged through the rushing clouds, and wonderfully dry air filled our lungs. What does a sailor do after a storm?

"Let's check on *BluBird*," I suggested to my wife.

Finding her in good condition, Susan and I set out for a sail, a wonderful respite that all avid recreational mariners understand. In between coming about and dodging other recreational boaters on a day sail from my eastern harbor to the western shore of Buzzards Bay and back again, Susan and I discussed how close we were to hitting pay dirt from our two-decades-old quest. I would have my answers at long last. My genetics came from Harvard Medical School. I was convinced of it! On that alumni list was John A. Williams; Roxanne's middle name was Williams, in honor of her eighth great-grandfather, Pilgrim Roger Williams. Could Dr. John be my ghost father? A quick iPhone search while on a starboard tack showed me that this Dr. Williams had died in 1990.

When I got back to the house a bit after 4 p.m., I had a voicemail from Roxanne and also an email in which she said that through the process of elimination, she had found a possible match—an uncle on her mother's side with the last name Wilcox. I felt the same rush that

I had experienced upon the discovery of my mother's secret over two decades earlier. I raced to grab the list of names, and there he was: Roger E. Wilcox. *There you are, you ghost!* The list said he had graduated from Harvard Medical School in 1948, and I immediately opened my laptop and googled him. I found his obituary; he had died in 2006 from complications from congestive heart failure. He had served in the US Army before graduating from medical school. He did his medical residency at Massachusetts General Hospital, and then served in the Army again during the Korean War, in the 8063 M.A.S.H. Apparently, he was quite a character and provided the inspiration for some of the stories used in the M*A*S*H book, movie, and TV series.[11]

Victory! Within twenty-four hours of first learning of an Ancestry.com cousin, I was sure I had found my biological father. Roger E. Wilcox's timelines all fit with the story my mother told me in 1995. I plugged the name Wilcox into Ancestry.com to see who appeared in my DNA connections. One name surfaced: Roxanne's! I quickly sent her an email and asked more about Roger.

Before I had any time to celebrate this new revelation about the identity of the ghost, Roxanne called and quickly took all the wind out of my sails. "You've got the wrong guy. His name's not Roger. It's Granville."

I had jumped to the wrong conclusion, and much too soon. "When you said 'Wilcox' and there he was on the Harvard Med School's graduates list, I thought for sure we nailed it," I told her.

This was some quick round trip—from a black, empty hole in the ground to the panoramic mountain top, and back again in the course of an hour. Bummer! *Wrong conclusion, right outcome on the horizon*, I thought. But I was so much closer to discovery than I had ever imagined I would be. I could taste it.

---

11 "Roger Eugene Wilcox," originally published in *The Arizona Republic*, August 2, 2006.

It turned out that the much older doctors in her family tree did not match my conception date, and they had no Harvard connection. I knew I was impulsive and told myself to hold back from taking a much too premature victory lap. But victory was close. "So Granville's not a doctor," I said to myself. "I can live with that!"

Roxanne explained why she believed her Uncle Granville was my ghost father. Her father was an only child, and his father, also an only child, died in 1942. Unless her paternal grandfather had a child out of wedlock, an unknown great-uncle, then maternal great-uncle Granville was the most likely candidate.

"I'll ask some of my Wilcox cousins to take the Ancestry.com DNA test too. But it will take about six to eight weeks."

What was six to eight weeks in the grand scheme of things? I thanked her for her continued efforts. Then, before she hung up, she said, "My husband and I are planning to visit our investment property in Provincetown at the end of September. Would you like to meet up in Hyannis at that time?"

I felt ecstatic to have discovered a new cousin and to be en route to making a new friend. I reflected on the emotional ride this connection had been for me—after twenty-plus years of attempting to find a missing link, to have made such progress within just a day and a half of the Ancestry.com introduction to Roxanne was incredible. I added to our conversation, "Feel free to share my personal info with your Wilcox cousins. I'm excited to meet you in Hyannis at the end of Sept." We decided to have coffee at a Starbucks on September 26 at 11 a.m.

Contained. Buttoned up. Controlled. Calculating. Clear-headed. All of these were appropriate terms for how I managed my emotions when under stress. No more! I felt uncharacteristically fragile and expressive as the prospect of identifying the ghost looked more and more like a genuine possibility.

In the days that followed Roxanne and I exchanged more than

seventy emails, sharing family pictures, childhood stories, adult hopes, dreams, disappointments, and achievements. Speaking of disappointments, the more I learned about Granville the ghost, the less connected I felt to him.

It seemed that Granville had lacked any interest in the family cotton brokerage business and had little early career direction. After serving as an Army medical specialist in WWI, Granville worked first as an electrician, then as a salesman, and finally as a general manager of a small manufacturing business. He had a son with his ex-wife, whom he never saw. She raised the boy, named Stanton, in Minnesota, without the involvement of his father or any of his father's family. Stanton grew up to become an accomplished Air Force lieutenant who, at age twenty-four, became a missing-in-action pilot when his Starfire reconnaissance plane was shot down over North Korea on May 3, 1953. He and his copilot were presumed dead one year and a day later.

Over Labor Day weekend, Roxanne informed me she had uncovered the Korean War Project, a DNA laboratory set up at Dover Air Force Base, Delaware, to test unidentified Korean War remains against the DNA of family members with relatives who had gone missing in the Korean War. *What a research machine*, I marveled. We both ordered kits in hopes of finding Stanton. If Granville were, in fact, my biological father, Stanton would be a DNA match, my half-brother. An hour and a half later, she then advised me about one of her Wilcox cousins.

At the end of one our numerous emails, Roxanne wrote:

My Wilcox cousin Llarisa will be contacting you by email this weekend! She's willing to do the ancestry.com DNA!!!!!!!!!! I shared some of your pictures with her.

Hurricane Irma had been brewing in the Caribbean over that Labor Day weekend, but Cape Cod's weather pattern had turned glorious for beach goers. The winds were a bit brisk for fair weather sailors: 25 knots out of the southwest. That spelled a choppy sea on Buzzards Bay waters. Perfect for a long, wavy beach weekend with two visiting grandsons and their parents! Since I had thought of little else other than Granville in the past four days, I welcomed a bit of a reprieve from all the conjecture without real data.

Over the next two weeks, Roxanne and I exchanged dozens of emails and had weekly phone calls. Most communiqués were informational. We talked about ourselves and our lives. The more we learned about each other, the more connections we started to find.

Roxanne was born in her father's hometown in Vermont. She went to grade school and high school near Pittsburgh, where her father gained employment as an engineer in a steel-related company; her family summered on Cape Cod. Her father had retired after her graduation from high school and moved full time to Cape Cod. She and her husband had graduated from my alma mater, the University of Massachusetts, Amherst, seven years after I did. My old haunts were also theirs.

We talked at length about rogue Uncle Granville. Why was a fifty-year-old man donating sperm? Answer: He was disinherited and needed some extra money. What was he doing in Harvard or Boston? Answer: He went to Boston regularly. And he knew people in the medical field. If Granville had donated sperm to my parents, why did Roxy show up in Ancestry.com as a "Close Family, 1st Cousin?" instead of a 2nd Cousin? Answer: The shuffle of the DNA deck works that way sometimes.

Roxanne commented that Granville visited Newport, Rhode Island, frequently, and I realized that my parents had lived in Newport around that same time frame. My mother's judgment was generally much less than stellar when she was under pressure. I gave my imagination permission to color outside the lines. Could she have concocted a Plan B if artificial insemination wasn't working? Could a quick tryst with

Granville have been her Plan B? Anything was possible, but learning the "who" was more important to me than the "how."

If Granville were, in fact, my biological father, I wanted to see where he lived in 1945. Ancestry.com's census data pinpointed his address. On September 14, after returning via Amtrak from a New York City business meeting to the Providence train station, I drove past that address, which was now part of an expanded Rhode Island School of Design.

The twenty-second anniversary of my mother's surgery, stroke, and unfolding secret was just a few weeks away. Llarisa's DNA results, at best, would take another five weeks. I was obsessed.

But Llarisa's results would be five days away, not five weeks. At 12:42 p.m. on September 19, just three weeks to the day of our first introduction, Roxanne sent me an urgent email that I did not notice:

CALL ME! I HAVE NEWS!

Twelve minutes later, my cell phone rang. "Where are you? Did you see my email?" she huffed, sounding somewhat impatient.

I had driven to Woods Hole. Tropical Storm Jose's rain was subsiding, and its winds had substantially weakened—no big deal. I had just finished a scrumptious bowl of kale soup as a luncheon treat from Pie in the Sky, a favorite local Portuguese bakery and luncheonette.

"No, not yet. What's up?"

Roxanne's voice had a cheerful chime to it. "Are you sitting down?"

"No, I'm walking in some fog and a light drizzle back to my car."

She continued, "Get a pen and some paper and write down what I'm going to tell you."

"That's inconvenient, Roxy," I replied. "My car is a ways away and I'm not sure if I have any paper in the glove compartment. If you've got something to tell me, go ahead. I can remember without writing it down."

Roxanne briefly outlined what she had discovered as she told me, "Something wasn't adding up for me. I called Ancestry.com's 800 support number, and spoke to a guy named Dave, who reviewed a bunch of technical terms with me. I found out something amazing."

I knew instantly why she had asked if I were sitting down. She had uncovered something, but what was it?

Before I had a chance to ask, she said, "We're not cousins. We're half-siblings. My father was the source of your sperm donation."

Jose's remaining drizzle momentarily stopped falling as I abruptly halted in my tracks. I felt so astonished that my legs were weak. I grabbed hold of a parking meter on Water Street, near the Woods Hole Ferry Terminal, to steady myself.

"How does this make you feel, my new big brother?" she asked.

"This is a home run, Roxy. Just knowing is a home run. You're a home run. And I'm beyond thrilled that YOU'RE the one who told ME."

"So, go home, call Dave, and then call me back," she concluded.

Chapter 20

As advertised, my oyster-gray all-wheel-drive sedan gripped a slippery road with fly-paper traction despite the rain, sleet, snow, wet leaves, and whatever else the storm had brought in. I raced through the roadway pools of Jose's tropical rain on the slick, curvy ride from Woods Hole to my swivel rocking chair. Fortunately, mine was the only vehicle on that winding back road. A drive that typically took twenty minutes took just seven minutes that day. No bikers, golfers, or playing children were out and about for that streak of oyster gray to run over by mistake.

Some of the crushed shells in my driveway were floating in Jose's pools as my sedan splashed to its resting place. I bounded ankle-deep in rainwater from the spongy driveway to my front door.

"Well, don't you look like the cat that just ate the canary? Take your wet shoes off, please," said my wife, who had been a full partner in a nearly twenty-two-year search.

Shoes could wait.

I sloshed into the dining room where she stood, hugged her, and breathlessly told her of Roxanne's phone call. What can anyone say?

Susan's was the third jaw that had dropped that day. Roxanne's was first. Mine was second. Susan's eyes teared up, and her smile turned to a beacon of light that cut through the ocean fog. We hugged and giggled until we could hear the driveway rainwater from the bottom of my pant leg dripping onto the dining room carpet.

"Wow! How about that! What a treat. Now you know."

I kicked off my soggy Sperry Top-Siders, grabbed the telephone receiver, sat down, and logged in to Ancestry.com on my laptop as I dialed the posted 800 customer support number.

"Hello, this is Terry at Ancestry.com. How can I help you?" a pleasant female voice asked on the other side of the line.

"Terry, please transfer me to Dave. I just got off the phone with a relative, seemingly a new sister, who had spoken to Dave. She asked me to log in to Ancestry.com, call Dave, and hear directly from him what he had told her."

"I'm sorry, but that's not how we work," said Terry very sweetly. "I can help you. Just log in and we'll go from there."

"I'm already there," I told her as I gave her a brief rendition of my sperm donor heritage and the interchange with Roxanne. It seemed that Terry had seen this movie before and had been well trained in performing her role in the story. This was my first screening.

Terry instructed me to click on DNA Matches, look directly at the connection between myself and Roxanne, and read it to her. I complied.

"What do the numbers read underneath that?" she asked. I had not paid any attention to those cryptic numbers earlier.

"Shared 2,035 cM across 61 DNA segments."

Groan! This again! I had no appetite for Google-searching new terms once again. Give me immediate gratification, please!

"Terry, I'm seventy-one years old. I graduated from college fifty years ago as a liberal arts major and a business minor. What the heck is a cM? And what's with the question mark after 'close relative, first cousin'?"

Terry was very calm and patient, with a helpful, friendly style. She gave me a light chuckle and explained. "A centimorgan (cM) is a measurement of how likely a segment of DNA is to recombine from one generation to the next. It was named by the discovering scientist, Alfred Sturtevant, to honor his teacher, Thomas Hunt Morgan, a famous American geneticist. The original unit was called a morgan, but that term isn't used today."

She paused before she continued. "When two people have a DNA match, it means they inherited DNA from one or more recent common ancestors. The length of DNA they have in common is estimated in centimorgans. The higher the number, the closer the relationship."

She continued, "Click on the icon that is a small, italicized letter 'i' in a shaded circle, just under those shared DNA numbers."

It was, I thought, the most unobvious icon ever placed on a web screen. Icons as important as this one should beg to be clicked. This one begged to be overlooked.

I clicked. Once I did, a three-columned table appeared. On the left, it showed percent of probability. In the center, a range of centimorgans. On the right, it listed the type of relationship.

Terry instructed, "Look at the list on the center column. Now scroll down to where your 2,035 centimorgans fall and read it to me."

I scrolled past Parent/Child, who shared on average 3,276 centimorgans, and Full Siblings, who shared a range of 2,400 to 2,800 centimorgans. I stopped at Close Family. It stated that there was a 100 percent certainty in a range of 1,450 to 2,050 centimorgans that the relationship was grandparent to grandchild, aunt or uncle to niece or nephew, or half-sibling. Just under the Close Family entry, I read that first cousins, half-aunts or -uncles, and half-nieces or -nephews shared a range of 575 to 1,300 centimorgans.

"Let's talk about the possibilities," said Terry, who now had become my best friend and closest confidant.

I repeated the sketch of my sperm donor origin. "I'm seventy-one. Roxanne is sixty-five. The grandparent to grandchild possibility is out. Given that her father was an only child, as far as we know, and her grandfather died three years before I was born, the aunt or uncle to niece or nephew relationship is likely out, too. And we were way above the first cousin/half-niece centimorgan range. The only thing left is half-sibling. We're at the high end of that range to boot—2,035 versus 2,050 centimorgans. That's within spitting distance to a full sibling, which begins at 2,400 centimorgans, but isn't quite there."

"Congratulations," she said. "You've just met your half-sister."

This was likely the same process that Roxanne experienced with Dave, I thought.

Terry continued after a brief pause, "The question mark is there because there are, as you just read, a few Close Family choices to select." She paused again, then sensitively offered, "DNA doesn't lie. Are you okay?"

You bet! For the second time that day, I could not see through my tears; nearly twenty-two years of digging, scratching, searching, and then pay dirt. "Not just okay, Terry. How about ecstatic?"

A few minutes after Terry and I hung up I called Roxanne. "Is this my new big brother, Morgan?" she asked.

Not a newly named hurricane from the 2017 season, but a derivative from our shared 2,035 centimorgans across 61 DNA segments, my endearing new nickname kicked off an equally endearing new relationship.

Being nicknamed "Morgan" provided a superbly richer, more rewarding aftertaste than any other success, which I celebrated with some dark chocolate and old cognac. The identity of the anonymous donor who had created me was buried inside Roxanne's family tree. She had persevered to find my biological father somewhere in her family tree, not realizing then that she would discover that her own father was also my biological father.

We now knew the "who." Roxanne's father, Donald, was the sperm donor for my parents. We both repeated the words of Dave and Terry: "DNA doesn't lie."

Over the next few days, our telephone and email connections were on fire. Roxanne filled me in on most everything one might want to know. She provided pictures of him at varying ages and described his personality quirks, education, occupation, hobbies, and his excellent health until his death in September of 1995 at the age of eighty-six. Roxanne unveiled her father's family tree, which contained not only Pilgrim and colonial leaders but also those with rank and privilege throughout old-world Europe.

My family compared Donald's school graduation picture with Stephen's picture at the same age. "Unquestionable! Meet your grandfather," exclaimed my daughter-in-law to my son, who had resembled no one in the family until that moment.

Stephen confided that he had experienced several emotions all at once. He was thrilled for me to have discovered my paternal origin after more than two decades of searching; happy for the reassurance of his own inherited good health; ambivalent about a new family heritage with which he had not established a childhood connection; and fearful that the discovery of a new close relative, an aunt, would also come with obligations that he was not prepared to fulfill.

Whoever that fertility doctor was, he found a sperm donor who was a fine match for my parents, I thought. Not Northern Italian, but almost a mirror image of my mother's British Isles and French ethnicity, except for the Scandinavian piece instead of her Ashkenazi Jewish piece.

Donald had graduated from a noted technical college with a degree in mechanical engineering. My dad was an aspiring engineer who had not completed his engineering education. Donald had an arc of blue in his eyes. Dad had crystal-blue eyes. While I inherited the blue eyes, that mechanical aptitude had passed me by. But I am reasonably intelligent and in excellent health. A generation later, that mechanical aptitude flourished in my daughter. And the apparent leadership gene landed in me.

But then the doubts crowded in. Almost in horror, I asked myself, "Was that fertility doctor a eugenics disciple? Am I a product of some sort of contrived genetic engineering science experiment?"

None of us have any control over how we are born. We just are. I just am. Not perfect, but no more imperfect than any other human being.

I heard my mother say, "Just do the best that you can with what you have."

During that period, I engaged in imaginary conversations with her; I was sometimes grateful, sometimes enraged. "How could you do this to me, conceive me in a way that kept all the knowledge about my genetic heritage from me?" But she did so knowing that this was the best practice for my well-being. I romanticized it to diffuse any residual anger; I was a loved and wanted child.

And, finally, I knew who my biological father was. Donald, with Mayflower and European royalty roots! *Pretty cool*, I thought. My ghost father was a Mayflower Yankee and not a deadbeat dad. I am not a Frankenstein monster. Roxanne unknowingly added some new fuel to my old biases and pangs of feeling flawed in a series of communiqués that not only welcomed, but also troubled me.

"Join the Mayflower Society and the Roger Williams Family Association with me!" she urged.

That struck a chord of fear in me.

Do not let this new sister be a silver spoon. A pretentious Yankee so enamored with her pedigreed family history and heritage that she, herself, has little to offer. Happily, I would soon find her to be just the opposite.

Even with the new knowledge that I shared her old-Yankee genetics, I still considered myself a logical Italian. That is how I was raised. I was proud of the first-generation American value system and work ethic that had been the foundation of my own upbringing; my value system was cemented permanently into place. No new genetic awareness could alter that foundation.

Roxanne was the bonus I had least expected. The sperm donor, the anonymous and healthy ghost father, was at long last revealed, thanks to a champion, my new sister who was not a cousin after all. What a fantastic bargain in the process of discovering the truth. I felt complete . . . almost.

There was more for us to uncover; so much more. How did a Vermont engineer donate his sperm to a Boston fertility doctor just three months before he wed? Who was that doctor? Where was there a record of a Dr. Sims in Boston on 10 Beacon Street? What was his Harvard Medical School connection? Did we share any more half-siblings from Donald's sperm donation?

Roxanne and I planned to meet for the first time on September 26. Thanks to Roxanne's intuition and corroborating research, a credible "how" scenario had emerged quickly. Jaws would drop even lower once we connected all the dots and completed our story. But no one was more surprised than we were.

Chapter 21

Tropical Storm Jose finally blew past southeastern Massachusetts and whirled its way up the North Atlantic, but the New England skies failed to fully clear. We were far too intoxicated and euphoric with the newly uncovered connection of paternal genetics to notice that Hurricanes Katia, Lee, and Maria had been brewing in rapid succession along the Caribbean.

In the midst of these newly named hurricanes, Roxanne's discovery of my paternity, the newly named Hurricane Morgan, had blown the roof off the house that held conception secrets so long hidden. Who in 1945 could have predicted that the progress of genetic science would toss the term "anonymous" onto the heap of obsolete promises? From that genetic road map, we found credible inferences upon which we built the "how."

My previous pace of research and discovery seemed paltry next to Roxanne's quick bursts of discovery. Our telephones and computers reverberated at the speed of light for the entire following week.

Even though my daughter had first found information about Dr. Rock and his fertility clinic four months prior, and I had discounted it, Roxanne connected some dots that we had not been aware of earlier. As she learned of Dr. John Charles Rock's obituary, his biography, and his fertility clinic's records from 1921 to 1985 tucked away in Harvard's Medical School library, she took her findings to a whole new level.

Within two days of discovering our Ancestry.com half-sibling connection, my new sister Roxy had nurtured a first-name relationship with and introduced me to Jessica, a professional at Harvard Medical School's Center for the History of Medicine. Roxanne learned that Dr. Rock's daughter had donated his fertility clinic records (which had focused on his research) to Harvard Medical School a year after his death in 1984. They had collected dust in boxes for over twenty years, until the library had digitized their contents in 2005. By that time, I had already moved to Philadelphia and had given continued research on artificial insemination a rest.

Jessica advised us to scroll through the titles of the various digital folders and highlight anything we might find revealing or interesting. Two folders, in particular, with the same title had caught both my eye and my imagination: "Vaginal Smear with Sperm to Fertilize an Egg," one dated February 26, 1945, and the other dated March 12, 1945. If Dr. Rock, and not Dr. Sims, was my parents' fertility physician, could either folder contain documentation about my conception?

My mother had always stated that my punctuality began at an early age. I was born "pretty much on time." Names in the files had been redacted. Any details were not to be revealed for eighty years, but Jessica vowed that she could gain some privacy exemption for us and investigate.

Meanwhile, I purchased several copies of Dr. Rock's biography from Amazon.com to give to Roxanne and members of my family. I asked all of them, "What we can learn from this story?"

We looked through the pages of the biography, nodded, and said, "Aha!" Dr. Rock kept two offices at the Free Hospital for Women in Brookline. His free fertility clinic was situated in the basement. His private fertility practice was headquartered on the third floor of a new wing at the rear of the hospital.

I could hear my mother saying, "Third floor, rear."

But still, pieces that directly connected Roxanne's father (and my biological father!) to Harvard remained missing. And what about my mother's clues? Dr. Sims from Harvard Medical School and his office in Boston at 10 Beacon Street? And where was that ad or article? Absent still.

The long-sought knowledge of "who" brought to the surface a whole new range of unfamiliar emotions to navigate as Roxanne and I prepared to meet face-to-face for the first time. I am not the nervous type. I have met with celebrities, political leaders, high-profile CEOs, billionaire investors, and famous advocates of varying types. But *this*? I was uncharacteristically anxious!

What to expect? How to act?

"Just be yourself," I said to myself. "Like it or not, just show her who you are; no more flawed than the rest of us."

What could I give to Roxanne that would be both fun and show my appreciation for her partnership and her deep dive as she dissected her family tree?

In collective wisdom sessions around the Cape Cod dining room table at sunset, I asked my family, "How about an engraved ax?" Roxy had joked earlier about googling me to ensure I was not an ax murderer.

They summarily rejected my sense of humor for a better option.

"How about a compass that's engraved with something like 'Glad you found me,'" my son-in-law offered. Creative lad!

Perfect! Hello Google! "Find compass gift!"

Inside of forty-eight hours, just the day before our introductory cup of coffee in Hyannis, I received a rush shipment—a gift box containing a working compass in the form of a silver necklace, engraved with the words, "LSR, Thx for finding me. BBP," on the back. Since we had both recently used the LSR (Little Sister Roxy) and BBP (Big Brother Peter) initials to address and sign a flurry of emails to one another, they seemed an affectionate way to express my appreciation on the engraving.

With a small gift box in one hand and a copy of *The Fertility Doctor: John Rock and the Reproductive Revolution* in the other, I asked Susan, "Would you drive? I'm too nervous to be an attentive driver this morning."

Chapter 22

On September 26, Susan drove us in our antique Buick convertible to meet Roxanne. The meeting spot was a Hyannis Starbucks not far from the Mid-Cape Highway. We arrived early. I wanted a few extra moments to stake the place out and grab the best seats in the house. The mid-morning coffee crowd on this somewhat overcast late September Tuesday was predictably thin. I claimed a vacant table at the back of a windowed wall facing the entrance.

Roxanne and I had confided to each other that we were both jittery and anxious about meeting as brother and sister for that first time. Would she like me? Would I like her? How should we greet each other? With a handshake?

The moment I saw her, impulse took over. We had exchanged pictures, so we both knew what the other looked like. I instinctively sprang from the table and greeted her at the door with a smile and a "Hello, new sister Roxy," followed by a hug and a kiss on the cheek.

Roxy oozed warmth. "What was that line in the Jerry McGuire movie? 'You had me at hello!'" Handshakes followed as we introduced our spouses.

Susan and Roxy's husband, Kaz, ordered our coffees while Roxy and I took our seats and embarked upon an hour of nonstop chatter. We sat side by side on a padded bench seat against the wall behind the rectangular table. After some brief small talk, I acknowledged, "You have no idea how overwhelmed I am at the efforts you put forth to find where I fit in your family tree. Not knowing where I came from has haunted me for over two decades."

While we had talked on the phone and over email, and she knew some things about me, I gave her more information about the circumstances of my childhood, including my dad's illness and suicide and its impact on me. "I'll be eternally grateful for the knowledge you've given me."

"Yes, it was quite a ride for me," she said with warmth. "I'm happy, too."

Once Kaz and Susan returned with our drinks, I presented Roxy with the Dr. Rock biography and the gift-wrapped silver compass necklace. Roxy teared up with obvious delight as she read the engraving on the back.

"I might have given you an engraved ax had not my family interfered," I joked.

"This will show much better," she gushed.

We talked for a while about the day that Roxy learned that we were siblings. Laughing, I said, "Yes, that took all of twelve minutes from your email to your cell phone call."

Kaz jumped into the conversation. "She vaulted through the front door, grabbed my hand, obviously overjoyed, and said, 'You'll never guess what I just learned!'"

Roxy chimed in. "Kaz followed our twists and turns for the entire three weeks. 'Bizarrely fascinating' is what he called it."

Kaz continued. "I couldn't believe it." He paused, then added, "I had known my father-in-law for nearly a quarter century." He turned his attention to me. "This would be just like him, to help a childless couple. Don was that kind of guy."

For the entire twenty-two years since learning that I was donor-conceived, I never once fantasized about meeting or having a relationship with my birth father. It was far more important to me to know his genetics and medical history. But after hearing Kaz speak about Don, for the first time I felt robbed that I had not had the opportunity to know him.

Our conversation flowed naturally and comfortably. We had known this couple for all of forty minutes, but it felt like we had been friends for the entire forty years they had been married.

"Should I reach out to my brother?" Roxy asked. She shared that she had a younger brother and that they had been estranged for years. But I was compelled to know as much as I could.

"I'm all through with secrets, Roxy," I said, taking on a serious tone. "Truth has set me free; so, yes, tell your brother. I don't expect anything. I don't want anything. I just want there to be no more secrets."

Following that, we discussed the several pieces of information that still did not add up. Why was a Vermont engineer in Boston donating sperm to a fertility clinic? How many times did he donate? Do we have other donor-siblings? Who was that twentieth-century Dr. Sims? Where was this unknown fertility office that we couldn't locate in Boston on 10 Beacon Street? What was the Harvard Medical School connection? How does Dr. John Charles Rock fit into the equation?

Before we departed Starbucks, we planned to meet for a second time in Boston a week later—in the lobby of the Back Bay's Lenox Hotel on what would have been my mother's ninety-seventh birthday and the twenty-second anniversary of the stroke that eventually let the cat out of the artificial bag. Before our dinner in Boston, Roxy and I planned to stop at the main branch of the Boston Public Library located right next door to the Lenox, for some research.

As we bid our farewells in the parking lot, Roxy said, "There has got to be a connection," she said resolutely, as she pointed as if aiming at an imaginary target, "and we're going to find it." Roxy enthusiastically exclaimed that she had some leads at Harvard University. "My father's best friend, Dwight, whom he'd met in college, managed Harvard's psychology lab and had befriended B. F. Skinner, the renowned Harvard behavioral scientist, during and after World War II. Dad visited Dwight and his wife, Shirley, in Cambridge many times while en route from Vermont to Providence when he and my mother were courting," said Roxy.

"She's a hoot," I said to Susan as she drove us home. I felt so light that I could have propelled myself home without the old Buick's help, like a hydrogen-filled balloon with a major air leak.

My phone and email were ablaze for the next few days as I filled my circle in on our coffee meeting.

"Do you have a picture?" most everyone asked.

"No! Jeez, we were so excited to meet that we forgot to take one," I told them all, receiving groans of disbelief.

Before meeting in Boston, Roxy and I exchanged a few emails. She said the same thing—that "We forgot" and "everyone groaned."

Having our picture taken together in Boston was near the top of our "to do" list. And, yes, she had called her brother. They caught up on the more than ten years that had passed since they had been in touch and, as she put it, "I clued him in on our three-week marathon discovery about our father's sperm donation, to include your identity and background. He asked, so I gave him your contact information."

Research at the Boston Public Library was dissatisfying, as I had suspected it would be. There was no trace of Dr. Sims in a fertility clinic at 10 Beacon Street in Boston; we found no 1944 advertisements or articles about a Boston fertility clinic in general.

The librarian's research assistant noted, "The topic of artificial insemination by donor seems to have been clandestine during that time frame."

I asked her about Liz and was disappointed to hear, "Sorry, she must have been here way before my time." But Roxy and I had fun asking the assistant to capture the moment. She more than graciously took our picture as we stood in front of some bookshelves.

We conversed candidly at a Cambridge restaurant during dinner. Roxy added another layer of genuine warmth and humor as she shared her conversation with her friends and family before our first meeting.

"Oh, please," she said humorously, "let him not be a staunch political conservative or an evangelical zealot!"

I laughed out loud and corrected her. "I proudly categorize myself as a political independent, fiscally conservative, socially liberal, and spiritual, not religious."

She chuckled, "Takes one to know one."

Our dinner conversation in a quiet far corner of the restaurant moved from light to deep. Roxy tenderly asked me, "How should I refer to him. I mean, Dad?"

I was far less tender in my response. Perhaps I was protecting myself. Just because I had undergone therapy, I had not become someone else; I was just more aware of myself. I could deny my feelings and think logically. Except this time, what I felt and what I thought were the same.

"Roxy, I had nearly twenty-two years to come to terms with my conception. I understand that this whole premise is new to you. Donald was your dad, not mine. It never was his intention to be my dad. He sold his sperm to enable someone else to be my dad."

She sat silently as I continued, "I'll be eternally grateful to him for the gift of life and to you for finding the source of that gift."

Just then, our waitress appeared, asked if everything was okay, and poured a refresh of Sauvignon Blanc into each of our glasses.

After the interruption, Roxy said, "I feel concerned that my brother hasn't contacted you. It sort of feels like a rejection, in a way."

Had I constructed another emotional defense? Gaining a personable sister from this whole search was too good to be true. I could not expect more, and I had not. "Not to worry," I said. "I now know what I needed to know. I did not pursue my origins to gain another family. I already have a family."

"I have something to ask you." Her voice quivered with vulnerability. "So, now you know who your donor is—do you still want me?"

Roxy added a perspective I selfishly had not anticipated. Up until then, this genetic discovery was all about me. What about her?

I was raised as an only child. My only close experience with grown siblings was observing how well my children related to one another into their adulthood, or how my wife related to her brother. They all had forged special friendships, fortified by their common heritage, memories, and family stories.

Roxy and I may have lacked the common memories and family stories, but we shared DNA and had forged a common bond. Silently calling myself a callous bastard, I did not hesitate. "For sure! We have a special bond. I'd love to develop our relationship."

She kissed me on the cheek and said, "This is good to know. I'm thrilled to have a brother in my life."

Roxy was far more than a special new friend. I felt sadness for her that she had not experienced that closeness with her own brother. We shared an unbelievable connection. I absolutely wanted to continue and grow that friendship.

At the end of the evening as we said good night, I commented to her as we hugged, "This is all so new to us both, Roxy. We'll figure it out as we go."

Roxy was a perpetual motion–research machine for several weeks thereafter as we both kept family and our closest friends apprised of the twists and turns of our journey. We each conducted story time like a serial detective show with an ever-increasing circle of friends and family. I shared the picture of the two of us taken at the Boston Public Library with a wider and wider circle with the same query to all. "Who does this look like?" More often than not, the answers were, "Obviously a family member," "A cousin," or "I didn't know you had a sister." More jaws dropped as we wove a rich surrounding history into our story. We loved it.

The question always arose, no matter our audience. "How did this sperm donation occur in the first place?"

Anyone who was there is dead now. We cannot ask them. But between us, we have pieced together a cogent enough story as to "how" using the information that we gathered at the Harvard Medical School's Center for the History of Medicine and a biography of a Harvard fertility doctor.

I always passed my new sister all the credit.

And Roxy kept digging. She called and spoke to one of the John Rock biography's co-authors, Margaret Marsh, a history professor at Rutgers. I met with her, too. Then Roxy searched and actually located Shirley, the ninety-three-year-old ex-wife of her father's best friend, who lived near Harvard Square at that time.

"Can you believe that Shirley was a practicing nurse at Cambridge Hospital?" Roxy reported after the first of her few calls with Shirley. "She wasn't all that well, but well enough that I could ask her some questions."

We'll never really know all the exact details for certain, but we have the general gist.

Chapter 23

To set the period of my conception in the broader context of history, by the end of 1944, Perry Como had completed his debut week hosting a new variety show, *The Chesterfield Supper Club*. Most everyone listened to their radios nonstop during those times. After WWII's D-Day, Franklin D. Roosevelt's Fireside Chats and war news consumed the airways. And so did casualty reports. Popular music or new detective programs like *Boston Blackie* provided welcome escape and relief from the grim war news. Most Americans experienced the heartbreak of war casualties: a relative, a friend, a neighbor. By the time I was conceived, General Eisenhower's Normandy invasion was six months old. A Pacific campaign nearly as dramatic as D-Day was just underway. The sweet taste of a hard-fought victory was in the air. It could have just as easily been the bitter taste of a hard-lost defeat. Support of Allied Forces throughout the Western world neared a peak.

While anxious Americans had devoured their newsreels and radio broadcasts throughout WWII, they were abuzz with bulletin reports in December 1944. One way or another, everything seemed to be on the line. My parents were well aware of the war news; they lived in Newport, Rhode Island, during the war, a Navy town and home of the US Naval War College. My dad was not draft eligible; he was forty years old and had no shinbone in his left leg. He supported the war effort as a tool and die-maker at the Newport Torpedo Station. My mother often reminisced about how she and Dad listened to their radio to hear Douglas Edwards reporting the news for CBS's *The World Today*.

I could relate to much of that context. Susan had given birth to our first child, Stephen, while we were living in an Army town, Fayetteville, North Carolina, for my special operations training at the US Army John F. Kennedy Special Warfare Center and School in Fort Bragg. I was thrilled to be a new dad of a healthy son and scared to death all at the same time. Our post-Tet Offensive world in 1969 was uncertain, too. The Vietnam War had different stakes than those of WWII at the end of 1944. My stakes at the time were *just* my life or death, not the American way of life. Still, I empathized with the wartime drama surrounding my parents' decision to attempt to have a long-wished-for child, and I could imagine my dad saying something along the lines of "We just have to live our lives," or my mother saying, "We're doing the best that we can."

Nerve-wracking? Desperate? Courageous? I could not find the right word to describe the feelings that these two people likely experienced while undergoing fertility treatment to conceive a child while WWII was coming to a climax.

The news they heard on the radio of their pale green 1939 Ford rumble-seat convertible was hardly conducive to an anxiety-free conception. In crescendo fashion, just as American bandleader Glenn Miller's plane had been declared missing over the English Channel while en route to Paris, all hell had broken loose. Hitler unleashed Operation

Autumn Mist, his daring and desperate Nazi offensive meant to divide Allied Forces, seize Antwerp, and temper Allied offensive operations. The Battle of the Bulge, the American counteroffensive, had just begun. At its outset, Nazi SS (Schutzstaffel) troops had reportedly brazenly executed eighty-four American prisoners. German forces had surrounded Bastogne. In response to their demands to surrender, Brigadier General Anthony McAuliffe, acting commander of the supply-challenged 101st Airborne, famously responded, "Nuts!" The term became an instant rallying cry and Allied morale booster that likely aided General Patton's progress in breaking the siege at Bastogne.

In the Pacific, nearly simultaneously, American troops, alongside their Filipino allies, landed at Mindoro Island, the Philippines, to honor General MacArthur's pledge, "I shall return." Kamikaze attacks were not the only hostile force. Mother Nature did not pick sides. Just three days after landing, newsman Douglas Edwards reported that Typhoon Cobra had capsized three destroyers of the Third Fleet, a tragedy that claimed eight hundred lives. Meanwhile, my parents were attempting to create a life.

My mother repeatedly said that in January 1945, their doctor (was he Dr. Sims or Dr. Rock, from Harvard Medical School?) gave my parents the dreaded news. My dad was irrevocably sterile. They had a choice to make: remain childless, adopt a child, or "semi-adopt" a child through artificial insemination by donor. *They made their decision quickly*, I thought. I had generally observed my healthy dad to be analytical and decisive. Artificial insemination by donor it would be. Did he feel guilty, flawed, resentful, or angry? As a child, I was never aware. He was just my affectionate dad. With one chance in six that the procedure would succeed, they had expected to take several more trips to their Boston fertility clinic over the coming year.

Roxanne's exceptionally creative and tenacious sleuthing provided dots of information from which we interpolated answers to our open questions of how, who, and where: How did Vermont-based Don come to donate his sperm to a Boston fertility clinic? Who was that Harvard fertility doctor (Sims or Rock)? And where was that mysteriously missing office on 10 Beacon Street in Boston? Roxanne's discoveries from speaking with Shirley, the Cambridge Hospital nurse married to Dwight (Don's best friend), fueled our subsequent discoveries.

I typed "Cambridge Hospital, MA history," into Google and learned it was one of the Harvard Medical School teaching hospitals; among its specialties were psychiatry and women's health. Connecting the fertility specialist dot to Shirley was certainly possible; the fifty-five-year-old Dr. John Charles Rock, with thick, meticulously combed gray hair, brown bushy eyebrows, and a movie star shine, was a Harvard Medical School professor of gynecology, a reproductive scientist, and a fertility specialist.

Roxanne's biggest connection led to a then-forty-year-old affable intellectual with horn-rimmed glasses, closely cropped brown hair, and fair complexion who preferred to be addressed by his initials instead of his given name, Burrhus Frederick. B. F. Skinner, the now-infamous behavioral scientist with a Harvard PhD, had been a visiting professor at Harvard during the two semesters of 1944–1945. While previously at the University of Minnesota, he had contracted with the Department of Defense (DoD) on what was termed the "Pigeon Project." Could a pigeon be trained to act as a guidance system to enemy targets? After all, Skinner had succeeded in training pigeons to play ping pong. Why not guide bombs? Apparently in mid-1944, the DoD opted to focus their research resources on a hush-hush project ongoing in Chicago to develop the nuclear bomb and cancelled the Pigeon Project. After Skinner's two 1944–1945 Harvard semesters as a visiting professor, he was preparing to relocate to a professorship at Indiana University. He later returned to Harvard to cement his storied career.

Shirley confided to Roxanne, "I babysat for the Skinners while I was a nurse at Cambridge Hospital."

Roxy reported, "She was evasive and somewhat cagey when answering my questions about Dr. Rock over a few different conversations."

I thought, *Omertà . . . still?* Roxy told me that she suspected Shirley had more to tell.

I assumed Shirley had been adhering to my mother's plan to take what she knew to her grave. Roxy planned a face-to-face luncheon with Shirley in hopes of gaining more insight. But other than sharing that Don and Dwight had started a chimney sweep business together to make some college money, she revealed nothing new.

To find the connection between Don, B. F. Skinner, and Dr. Rock (with a unique fertility specialty), Roxanne recalled stories that her father had told about his best friend, Dwight. They had enjoyed a lifelong friendship that began when they were first-year students in college in Worcester. After the war began, Vermont-based Don made several trips to visit his best friend, who resided with his wife in a three-room, second-story flat located in an old Cambridge neighborhood adjacent to Harvard University. Dwight's living room sofa had become Don's stopping-off point on his frequent trips to or from Providence to visit his fiancée and her affluent family. The two were to be married in June 1945. Shirley had recalled to Roxy how she had looked forward to attending what she had expected to be quite an extravagant wedding.

Given these connections, it was no coincidence that Dwight took a job, part time at first, full time later, to manage Harvard's psychology lab, working under B. F. Skinner. Roxy surmised that Skinner and Rock were close acquaintances. They were Harvard scientists and colleagues, and both taught an occasional class at Shirley's workplace, Cambridge Hospital.

Exactly how Don was recruited and became a Boston area sperm donor will likely remain a mystery. In discussions with Roxy, Shirley did not remember how or when Don met B. F. Skinner. Was it at a Harvard University party that they had attended, with visiting Don as an invitee?

Or at some other event attended by other scientists? Had Don been introduced to Dr. Rock by Skinner at one of those events? Or, as a wild card possibility, Tracy speculated that Shirley might have introduced the notion of sperm donation to her husband, Dwight, and/or his best friend, Don, to earn an extra buck.

While we do not know how or where Don and Dr. Rock were introduced, I have conjured up believable enough scenarios, perhaps not exactly correct, but in all likelihood directionally correct. In my research, I learned that during the winter semester of 1944–1945, prior to my conception, two significant social events took place. The first was a celebration of women in medicine, likely an elaborate, celebratory event honoring two high-achieving Harvard women and attracting a broad attendance, perhaps staff from all the Harvard teaching hospitals, including Cambridge Hospital. The second celebrated the accomplishments of Dr. John Charles Rock himself. He, along with his research partner Arthur Hertig and technician Miriam Menkin, were the very first to have fertilized a human egg outside the womb, creating the concept of IVF. Certainly, B. F. Skinner would have attended that celebratory event. Would Dwight and Shirley have attended? Did Don attend? Is this where his introduction to Dr. Rock took place? I posited a scenario to Roxy based upon the facts we knew.

"Picture B. F. Skinner standing in a small group with Dwight and Don. He had just introduced Dr. Rock to the duo as small talk began. Skinner maybe asked the couple, 'Do you and Shirley have any plans for children?' And Dwight perhaps replied, 'Not right now. Perhaps after the war once I begin my own teaching career.' Then B. F. would have turned his attention to his colleague, the accomplished reproductive scientist Dr. Rock. 'You and Shirley had better stay away from Dr. John, then, or she'll be pregnant in a heartbeat!'"

I continued with my imagined scenario. "Don and Dwight might have looked at Dr. Rock with befuddled amusement as he clarified B. F.'s remark. 'B. F. is on his second martini. We have attempted to alter his behavior, but to no avail.' He perhaps even chuckled, going on to explain, 'I'm a teaching gynecologist and medical scientist at Harvard Med School with what some call a new-fangled fertility practice that I run out of another of the Harvard teaching hospitals, the Free Hospital for Women. My fertility practice isn't all that new, but it is unique. Couples come to me if they're having difficulty conceiving, although I've seen more than my share of women with several kids who want help to not conceive.' He probably emphasized the 'not.'"

Roxy and I began another three-week research marathon. Looking at the actual history of the times helped add some known facts to support the credibility of our reconstructions of past events.

I was likely conceived in early to mid-March 1945 via a "vaginal smear with (Don's) sperm to fertilize an egg." My search for information prompted me to again examine historical facts. In the back half of the February 26 issue of *Time*, the short article written as a case study must have given any would-be sperm donor serious pause. "Artificial Bastards?"

US Circuit Judge Michael Feinberg in Chicago had just ruled that artificial insemination by donor was a legal form of adultery that justified divorce. Would any donor want to be caught up in that controversy? At that point in time, Don would be getting married in less than four months. If he were recruited to be a sperm donor, would he want to talk to Dr. Rock immediately after seeing that article? What might his pre-sperm-donation mindset have been? Would he really have been willing to help a physician commit "adultery by doctor" to conceive a bastard child? (That is the wording used by the legal community in the article in *Time* magazine.) If Don had an arrangement with Dr. Rock, the doctor likely had a monumental sales job ahead of him. Would Don have considered backing out?

In my estimation, Dr. Rock assured Don that he adhered to the strictest standard of care, the real-world practice that included enabling a child to be "legal." His or her birth certificate was guaranteed to be signed and attested to by the delivering obstetrician and would name the mother's husband as the child's father. Only Dr. Rock, his assistant, the mother, the mother's husband, and presumably Don as a donor would have knowledge of the sperm donation. Anonymity and the child's legal status were guaranteed; perhaps that gave Don the confidence he needed to give of himself to make me possible.

Of course, I was still left with my big missing piece. My mother said it had been Dr. Sims at 10 Beacon Street in Boston, not Dr. Rock at a Brookline clinic. All the dots connected once I delved deeper into the Free Hospital for Women—I found information that gave me an "aha!" moment, information I am not sure how I had missed before.

According to *The Fertility Doctor*, Dr. Rock's clinic was reported to have two offices inside the hospital. The "free clinic" for patients who could not afford a doctor's fee was on the bottom floor, the basement. Other patients visited his private clinic located on the third floor, in a new wing at the rear. My mother had mentioned over the years that they had had the means to afford the fertility clinic's fees; hence, her references to the third-floor rear office. It was there that she and my dad met with Dr. Rock, and the procedure using Don's donation took place.

Unbeknownst to Don, on the morning of December 12, 1945, a twenty-five-year-old married woman gave birth to his donor-conceived child. Aside from the child's blond hair and fair complexion, friends and family all exclaimed, "Ah, he looks just like his father!" The couple named their son after his Italian immigrant grandfather who had passed away that June.

Now, having finally connected all of the complicated dots, I called out to Susan, who I thought was in the next room. She was not home. I called her cell phone. It went directly to voice mail. I left a message: "Call me. You won't believe this!"

Within the next two minutes, which felt like two hours, I telephoned Tracy at work. I never call my offspring at their workplaces. It was just after lunchtime and I hoped she would be at her desk.

She answered with alarm in her tone. "What's wrong, Dad?"

Wrong? Nothing was wrong; everything was right. "Tracy, the address of Dr. Rock's clinic inside the Free Hospital for Women was 10 Beacon Street, not in Boston, but in neighboring Brookline, about a hundred yards from the line between Brookline and Boston. There were plaques of Dr. Sims, the father of modern gynecology, and the hospital's founder, Dr. Baker, in the lobby. My mother passed those statues each time she visited Dr. Rock's office on the third floor, rear. Chrissake!" After her squeals and wows, I outlined more details.

She said, "You did it, Dad."

"No, Tracy, *we* did it."

When I hung up the phone, I cried loud and hard, much like my mother had done when she had shared her secret twenty-two years earlier. The multi-year marathon was over, with a sprinted dash to the finish line. I was jubilant, sure, but I felt so fatigued. I declared victory and took a half-hour nap.

Don had passed away on September 22, 1995. He went to his grave confident that he had taken his secret with him. But just ten days after his death, on October 2, 1995, the only other living person who also carried that secret, the mother of the resulting child from his anonymous sperm donation, suffered a post-surgical stroke, and the old gates that zealously guarded her secret no longer worked.

Neither fertility pioneer Dr. John Charles Rock nor any of his patients could ever have fathomed that the advancing science of DNA analysis, a new phenomenon called "the internet," and a more permissive society would have unabashedly shattered Dr. Rock's presumed-eternal promise of anonymity within two to three decades after they, the keepers of that secret, were dead and buried.

Roxanne had recalled her mother saying that she and her father had difficulty conceiving and visited a fertility clinic for help. We have speculated and surmised that her father ultimately returned to Dr. Rock's Brookline office several years later, this time at his private clinic as a patient, of all things.

After six years of marriage and no sign of a pregnancy, Don knew they had an issue with fertility. Where to turn? Dr. Rock! On one of those visits to see Dwight, Don and his thirty-eight-year-old wife most likely visited Dr. Rock's fertility clinic on 10 Beacon Street in Brookline. The diagnosis—an irregular cycle and ovulation, somewhat due to premenopause and easily addressed. Roxanne and I invented their conversation.

"How did you know about Dr. Rock?" asked his wife.

"I met him at Harvard, through Dwight and B. F. Skinner."

The first of their two children was born in 1952, a daughter they named Roxanne.

We do not have to speculate to see that advances in science enabled two tenacious, creative people with shared DNA to uncover the truth.

Roxy and I met face-to-face again at the end of October while Susan and I were en route to our Philadelphia home for the "off season." We both felt empowered by all we had learned. Roxy had run a fact-seeking marathon over those two months, from late August to late October, and won the blue ribbon. I felt like I had run one long, slow, plodding marathon (twenty-two years) and then sprinted through two successive three-week Boston marathons, from Hopkinton to Boston and then back again.

Old secrets had become artificial secrets. Roxanne and I are eternally connected by 2,035 centimorgans across 61 DNA segments, and there's no secret about it; our relationship is not the least bit artificial.

The only remaining unsolved mystery is this: What article or advertisement in what unknown 1944 publication featured Dr. Rock's fertility practice? I can live with not knowing that piece of the story.

But our story lives on. Do we have more siblings yet to be discovered? How many?

Chapter 24

There seems to be no end to this revelation.

From 2010 to 2020, DNA analysis as an industry has vaulted from $1 billion to $5 billion in revenue generated. The business valuation of industry leader Ancestry.com soared to $4.7 billion when it was acquired in August 2020 by private equity behemoth the Blackstone Group. Several months afterward, Ancestry.com announced the appointment of a new CEO, a freshly recruited former senior executive at Facebook.

The collective revenue of the firms that make the DNA sequencing instruments that enable that analysis has grown from $1 billion to $3 billion over the same time frame. Equipment growth had shown signs of plateauing, but technological innovation has continued. For instance, a portable DNA sequencing device identified Osama bin Laden.

As innovation continues, I forecast that DNA sequencing will move beyond resolving genetic curiosity and begin to delve deeply into personalized medicine. Adding support to that thesis is none other than 23andme, which in February 2021 announced a public stock transaction with a $3.5 billion corporate valuation and a war chest totaling over $750 million in cash to enable it to focus on the healthcare aspects of DNA analysis. Major shareholders include global entrepreneur Sir Richard Branson and Fidelity Investments.

Where do I think this is heading?

Let's imagine a world just a decade into the future. I picture myself taking a nostalgic tour of my old hometown and shopping in what had once been the former pharmacy where I scooped and poured behind the soda fountain. There's a specialty counter where cigarettes were once sold; now, medical marijuana has taken their place. In addition, endless shelves store what seems like every medication known to humanity. I grab a package of MDX-Gold, which in 2030 is a readily available over-the-counter CRISPR medication to relieve my allergic reaction to mold and mold pollen (and I do hope for such a medication).

At the rear of the pharmacy, where the old soda fountain once prominently stood, I notice two six-foot-tall young men in the midst of what appears to be a serious, rather raucous and animated discussion.

The older brother, who I learn is Leigh, has thick, closely cropped, straight black hair and darker southern European features. He is home from college for a long weekend. The younger brother, Maurice, is broader and decidedly more muscular. He has a relatively fair complexion and longer, wavy, medium brown hair and is awaiting admissions news about his college applications. The brothers, both athletes, are looking to maximize their athletic performance using legal over-the-counter vitamins. Over the last several years, thanks to the help of credit card–operated pharmacy kiosks, the repurposed DNA analysis industry emerged as a high-performance growth engine, which attracted the market's mainstream buyers.

In the place of the soda fountain's counter and stools are two portable DNA kiosks, each the size of an end table, with a video display.

Enter a valid credit card, spit in a vial, and insert it. Then, poof! The user is sent an instant email detailing his or her DNA analysis. While the results include all of the standard information regarding ethnicity, centimorgans, and DNA segments of commonality within its database of other customers, the test now largely targets personalized medical applications. Based on a user's DNA, the analysis recommends the vitamins or CRISPR medications best suited to the individual's unique DNA and bodily protein content.

In my imagined scenario, supported by decades of research and my own personal experience, it does not seem farfetched to me that technology will lead us to this place. I hope that full disclosure will become the new de facto practice standard used by the reproductive industry and adoption agencies—that all details about the genetic identity of egg, sperm, and embryo donors will be shared. But I know that families will continue to keep secrets, nonetheless.

Coming back to the brothers in my imagined scenario, as they review their results, sent directly to their smart phones, they read some surprising news. First, unsurprisingly, as their mother had told them, she is a combination of Northern Italian, French, and Irish. Leigh's father was Southern Italian and Portuguese, but—and here is the surprise—oddly, Maurice's readout cites his paternal Polish and Czechoslovakian genetics, an ancestry that is quite uncommon on Cape Cod. And why did the brothers share nearly 2,000 centimorgans, rather than closer to 3,000 centimorgans, which would be more likely for siblings who share the same parents?

Both boys seem to understand all too well what is going on. Maurice pushes the face of his wristwatch. An interactive screen as large as a dollar bill emerges. His mother, an attractive, middle-aged brunette whose effervescence belies her age, pops up on the screen. "Mom, when is Dad coming home?" he asks her.

"Not for a while yet. Why?" she replies.

"Leigh and I are on our way. We have to talk," he says. Those are the same four words I spoke to my own mother at the Sunday dinner table in December 1995: "We have to talk."

Advances in science and technology come with unintended consequences. Secrets, especially family secrets, are artificial. I had not realized at the time that when my mother revealed her secret in 1995 that it was on Popeye the Sailor's sixty-fifth birthday. By 2030, Popeye will have just celebrated his centennial. The unveiling of my mother's secret was traumatic for me, but the truth set me free. I created for myself a much better handle with which to carry the baggage of my childhood.

Now that I have the benefit of full knowledge about my genetics and my donor, and I have developed a deep friendship with Roxy, my new sister, I am totally at peace with my personal history. Still, there will always be a nagging curiosity about the possibility that I have other close relatives. Was I the product of a targeted donation? Did Don donate his sperm to more than one couple? Do we have any other half-siblings?

The existence of other half-siblings looked unlikely—until the third anniversary of Roxanne's first phone call to me.

"Are you sitting down?" she asked.

Jane, age forty, had popped up on Ancestry.com as a paternal first cousin, with 847 centimorgans in common with me and 1,055 centimorgans in common with Roxanne. For Jane to be a first cousin to both of us, Donor Don needed to have had a sibling. But Roxy's father had been an only child—unless her paternal grandfather had a misattributed child somewhere, Jane was our "half-niece." One of Jane's parents, in our age range, was likely totally unaware that he or she was a product of Donor Don and our half-sibling.

Roxy lamented, "The poor dear. Her family will find all this new DNA information so traumatic. Let's not approach her. Let her approach one of us."

I quickly agreed, not realizing how I might feel about leaving this new connection to chance for the long term. I felt like my sailboat,

BluBird, heading out to sea with no wind, adrift and directionless. I kept repeating to myself, "No more secrets." If I had another half-sibling or more out there, I wanted to know it, and I wanted them to know it, too. And I wanted to avoid any risk that my own grandchildren might one day unknowingly and unwittingly fall for a paternal relative.

After a month, I was unable to resist the impulse any longer. I constructed a cryptic email via Ancestry.com to Jane telling her that I had been conceived with the help of a sperm donor. I shared what I had learned and suggested she might possibly be my half-niece. I ended the email asking if she would care to explore our connection.

Roxy was not all that happy with me when I related what I had done. "How subtle! This should be interesting," she quipped. After a pause and a sigh, she continued. "My life is full. I know what I need to know. You're all I need. I understand your view of 'no more secrets,' but you're on your own on this one."

With all of the media publicity regarding donor-conception over the past several years, I had hoped that I would find other donor-conceived people like me. People who could relate to the range of emotions generated by the prospect of discovering more donor-siblings and who could give me some advice or lend a sympathetic ear to the findings and how best to deal with them. I had felt alone for the first twenty-two years after my mother's revelation that I was donor-conceived. But for the past three years, I had not felt so alone now that I had my new sister. My past feelings of isolation had seriously faded, but now it appeared I was on my own again.

Since my conception had not originated from a frozen sperm bank and my parents' fertility specialist was not the unethical donor, I did not expect to find a hundred half-siblings, if any existed at all. But I did ask: "Do I have another sibling?" I just wanted to know; I did not need to create another family.

I googled "Donor Conceived and Misattributed People," which

listed several social media support groups. With DNA Detectives prominently featured, the one that seemed best suited to my purposes was another private Facebook group, We Are Donor Conceived. When I clicked, I was asked to provide details about my situation, which would be reviewed by the group's monitors before I could be accepted. I hesitated at first, fearing this may have been a social media scam rather than a legitimate support group.

Doing more digging before signing up, I discovered that Bill C., the one who had uncovered that his parents' fertility doctor had used his own "anonymous" sperm, was also a member. From his own digging, Bill found more than sixty siblings, all from the same community, many of whom had no idea that they were donor-conceived in the first place.

We Are Donor Conceived had only been in existence since 2016 and included several thousand people around the world who had experienced the impact of misattribution and genealogical bewilderment from the surreptitious practice of artificial insemination by donor. Everyone had their own unique story. Like me, they had experienced a range of emotions, sometimes simultaneously: anger, relief, violation, deceit, curiosity, shock, shame, isolation, numbness, pride, grief, confusion, embarrassment, emptiness, sadness, joy, fulfillment, indifference, or a combination of high and low feelings that changed over time with more knowledge. Members of the group shared how they had discovered, processed, and benefited (or not) from what they had discovered. I was no longer alone; I had a non-judgmental community with whom to share feelings, tactics, and strategy.

Too bad I had not sought their guidance sooner. Once I brought the Jane situation to the group (after the fact), their collective wisdom related to me that my approach to Jane might have scared her off. Better to have communicated to her that I would be available to explore how we were connected and ease into it from there. They also warned me that, even with a perfect communiqué, she might not respond. Since I could not take the email back, my best hope was that Jane would ask her parents to take the Ancestry.com test.

Just three months later, Bernadette appeared on Ancestry's site as another "close relative" (1,755 cM in common across 41 DNA segments). (Her appearance on Ancestry.com was likely a product of the company's most recent Black Friday promotion, which attracted 1.5 million new customers.) Through social media, Tracy, Roxy, and I scoped her out immediately. Bernadette had been born in Springfield, Massachusetts, in 1946; she was an only child and most likely another half-sibling. *Poor soul*, we thought, *clueless and confused by the report*. I messaged her as my support group had suggested to "explore how we're connected." With some definitive knowledge, perhaps more truth will unfold, and she will respond to the overture. Those chips will just have to fall . . . or not.

Meanwhile, with new siblings occasionally popping up, I have to recover from feeling not quite as special as I did before these new siblings appeared. I'm not one of multiple products of a frozen sperm bank or an unscrupulous fertility doctor, but I no longer ask, "Are there any more siblings out there?" Rather, I ask, "How many more siblings are out there?" I felt a bit commoditized . . . not quite as special . . . after discovering just two others. How might I have felt after uncovering several dozen or one hundred? Members of We Are Donor Conceived encouraged me to download my raw DNA analysis from either 23andme or Ancestry.com and upload it to GEDmatch, which amalgamates varying DNA sites to give a comparative window into the universe of DNA customers. Law enforcement embraced this tool in 2018 to identify the Golden State Killer. One of my messages read, "If you have more siblings out there, this will surely help find them." No more secrets!

Could my curiosity be better directed? Yes and no.

I have come to learn that some of the younger, sperm bank generation of donor-conceived people have achieved earlier sibling contact success on the Donor Sibling Registry, founded in 2000 by a donor-conceived son and his mother. Professing a global presence and 70,000 members, the Donor Sibling Registry claims to have connected nearly 20,000 siblings—a great success for donor-siblings seeking each other. Followers are unsure how viable a pathway such a registry will continue to be in the age of low-cost DNA analysis over the internet. In my case, it is unlikely that any sibling I have even realizes that he or she was donor-conceived. I would likely only be delivering a traumatic, rather unwelcome surprise. There is no handbook written by the reproductive industry to help the donor-conceived deal with these revelations.

The century-old Seymour and Koerner handbook on artificial insemination by donor, equipped with the old practices of secrecy and anonymity, is now obsolete. Today, there is no place to run and little reason to hide. Society and its norms have changed. The American Society for Reproductive Medicine, which still publishes its regular newsletter, *Fertility and Sterility*, has emerged as the closest replica of a US governing authority of the reproductive industry, without the actual authority. Rather, it recommends a series of practice standards to be followed by its fertility practitioners and sperm and egg bank members located in more than a hundred countries.

While the American Society for Reproductive Medicine strongly encourages disclosing to the child the genetic details of his or her donor, including the identities of the donor and any siblings, and suggests limitations on the number of times the same sperm or egg cells should be used to conceive, there are no universal reproductive industry practice requirements. Other countries throughout Europe and the Pacific regions have enacted specific laws mandating restrictions on the repeated use of donor sperm, eggs, or embryos; required release of non-identifying donor specifics; and restrictions on donor anonymity when a donor-conceived individual reaches the age of eighteen, should they request information about their donor. Governing authorities, such as the UK's Human Fertilisation and Embryology Authority, have been established in several countries to oversee and enforce these laws. Yet, in several other countries (Spain, for instance), only anonymous donors are legal, and their selection is mandated based upon blood type and other genetic physical attributes of prospective parents. Others (Italy, as an example) have flip-flopped between treating reproductive assistance as unlawful or authorized with strict practice restrictions. Throughout the Middle East, donor-insemination has remained flat-out illegal.

Industries left to police themselves have generally failed to provide strict standards and rein in abuses. Hence, the US has a plethora of rules, regulations, and organizations to provide proper oversight, such as the FDA (Food and Drug Administration), the FTC (Federal Trade Commission), the SEC (Securities and Exchange Commission), and the FCC (Federal Communications Commission).

The Uniform Anatomical Gift Act, drafted in 1968, and the 1984 National Organ Transplant Act have established standards for organ and tissue donation in the US. Yet, there is no such governing body establishing laws and standards in the life-giving multi-billion-dollar reproduction industry, which has a history of ethics violations and abuse. Reproduction laws that cover donor-conception vary by state. In 2011, Washington became the first state to legislate open donation unless a donor specifically requests anonymity. The entire topic remains under-addressed by most states' laws and totally unaddressed by US federal law.

Professional fertility specialists generally belong to the Society of Assisted Reproductive Technology (SART), which inspects IVF clinics regularly and publishes IVF outcomes and success rates in transparent fashion. By 1988, the Clinical Laboratory Improvement Amendment (CLIA) required all clinical laboratory testing on humans in the United States to adhere to stringent standards. With fertility practitioners seemingly governed, despite the high profile of an unethical minority,

I'm struck by two issues. First, a guideline is not enforceable; a law is. Second, the free market distribution of not only sperm but also egg and embryo enables a multitude of unknowing siblings to be conceived from the same donor.

Without a specific legal framework, sperm, egg, and embryo banks have constructed their own sets of rules and regulations that look like a patchwork quilt when compared to one another. Some offer multiple levels of privacy that vary from providing information on the donor's identity to the donor-conceived child once the child reaches age eighteen (if they want to have the information) to anonymous options where only the physical, ethnic, and health information of the donor is provided and no information about the donor's identity is released. Once again, now that DNA analysis is so easily accessible over the internet, not only do I not get the lack of laws and regulations, but my sense of right and wrong continues to feel violated by the absence of relevant law.

There is more legislation protecting unborn dogs and puppy breeding than there is for the entire reproductive industry.

With no lawful regulations on the number of children who can be conceived from any one donor, the dozens (or even hundreds) of half-siblings have no protocol to ensure they are aware of each other and to prevent any sort of inadvertent incestuous relationship. Also missing are standards for the proper degree of genetic or medical screening of donors, which could prevent the transmission of inherited physical and psychiatric illnesses. And finally, we lack specific governance that punishes doctors for unethically using their own sperm multiple times to impregnate their fertility patients.

Psychologists have concurred in multiple studies that early disclosure is best for the longer-term emotional development of donor-conceived children, just as has been the updated practice for adopted children. Yet, donor anonymity has been left to the varied interpretations of parents, donors, and gamete banks.

The United Nations Convention on the Rights of the Child in 1989 (the United States was the only major nation not participating) and the European Convention on Human Rights in 1998 highlighted children's basic rights: "The child . . . shall have the right to know and be cared for by his or her parents. . . ."[12] The conventions focused on protecting children from forced military conscription and sexual trafficking. The visibility and acceptability of artificial insemination by donor became far more prevalent in the twenty-first century; yet, despite lobbying, those genetic rights remain unaddressed. Legislators have given all those rights thus far to donors and recipients. The donor-conceived remain on an isolated island, even though professionals have all agreed that their health and psychological well-being are at risk.

We Are Donor Conceived has produced its own pro forma recommendations of best practices for the donor-conceived. The recommendations are the product of multiple surveys of its membership from fifteen different countries in North America, Europe, and the Pacific. Anonymous donors, with no identity disclosure agreements, accounted for 89 percent of the group's origins; 4 percent discovered that their parents' fertility doctors quietly used their own sperm. The overwhelming majority of donor-conceived children who were surveyed believe that the assisted reproductive industry as a whole "has a moral responsibility to act in the best interest of the people it helps create."[13]

The older ones (people like me) were generally traumatized by a late discovery and often described their feelings as distressed, angry, or sad. The younger ones, between the ages of twenty and forty, making up 87 percent of those surveyed, were more likely to have learned of

12 Convention on the Rights of the Child, United Nations Human Rights Office of the High Commissioner, https://www.ohchr.org/en/professionalinterest/pages/crc.aspx.

13 *2020 We Are Donor Conceived Survey Report*, We Are Donor Conceived, posted September 17, 2020, https://www.wearedonorconceived. com/2020-survey-top/2020-we-are-donor-conceived-survey/.

their donor-conception as children or teenagers from their parents, as opposed to by accident well into adulthood (like me) through the surprising results of a DNA test or vindictively from an angry relative.

While having the knowledge at a younger age muted their feelings of trauma, it did not diffuse their natural curiosity about their genetic origins and their underlying medical heredity. According to the group survey, relationships with biological relatives was a pressing issue: DNA testing had identified biological parents or half-siblings for the vast majority of donor-conceived people (78 percent and 70 percent, respectively).[14]

My unknown origin consumed me for a quarter-century. The shock of my discovery has long passed. I've educated myself about the history of my donor-conception. My gene pool, equipped with names and faces of real people, is completely revealed. I feel complete.

Every now and again, like cracks of vibrating thunder, something triggers my emotions to continue a rollercoaster of a ride that I imagine will last a lifetime. *How many more are there like me?*

---

14 *2020 We Are Donor Conceived Survey Report.*

Epilogue

The biggest void I continue to feel is the absence of sitting down with my parents to highlight all the hidden treasures I have uncovered. After all my years of research and self-analysis, aided by my genetic discoveries via what was totally unforeseen in 1945—DNA testing over the internet—how might they have reacted?

I recently visited St. Patrick's cemetery, where my parents are buried, for an imaginary conversation with them. Before heading to the site, I reached for a compact fishing tackle box jammed in the rear of my old Buick's trunk. I opened the box and removed an antique fishing knife. I had long since tightened and sharpened its once dull, loose blade and polished its handle. It was the same knife I had allowed my mother to discover in my pocket when I was a twelve-year-old eighth-grader contriving our return to Cape Cod from Chicago. Dad's old keepsake symbolized my approach in life. When in a tough spot, with circumstances outside of my control, I tend to formulate a pragmatic game plan, marshal some available resources, seize something within my control, and make a bold move. I am the Go-to Guy.

A gentle breeze rustled the bright, newly sprouted spring oak leaves and wild pink beach roses that surround the family plot. The family monument overlooks the lily pads on top of Mill Pond. I recalled fishing from shore with my dad over six decades earlier after paying a Memorial Day visit to his family gravesite.

"Such a tranquil spot," I said out loud.

My time is approaching. I am not immortal. My children ask me on occasion to declare where I want to be laid to rest. I continue to wrestle with the notion of lying in the Guidaboni plot for all time or making other arrangements. Yes, I am a logical Guidaboni, even if I am semi-adopted. Is this where I belong?

On this visit, I had intended to trim the invading grass that covered my parents' footstones. As I cut away the turf, my hands and my dad's polished and restored ten-carat gold wedding band on my left ring finger were covered with earthen grime from the rich, wet soil. Dad had been resting peacefully in his coffin below for almost sixty years. I could feel him as he looked up and stated quietly, yet proudly, "My son is the most ingenious man on the planet."

Although he had limited time on this earth, I imagined that he felt at peace, assured that he had accomplished his ultimate mission as a dad; he taught his son to create a fulcrum for the heavy lifting in life. That ring shone through the muck on the world's strongest set of hands.

Appendix:
Additional Research and Insights

While my journey to discover where I had come from, and perhaps who I am, came to a relative end with the meeting and subsequent friendship with Roxanne, my curiosity about artificial insemination by donor and how the field had grown since my own origin had not. When the first wave of my artificial insemination research began in late 1995, the internet was in its infancy and Google did not exist. By the end of 2017, the internet flourished, and Google was a tried-and-true partner assisting me in the second wave of my research. I found that our society, culture, and science had begun at long last to move in step, although our culture required great leaps to catch up to the centuries-old head start held by assisted reproductive technology.

Both Roxanne and Tracy helped me determine that Harvard Medical School's Dr. John Charles Rock was the likely physician who used Donald's donated sperm in an artificial insemination procedure during the late winter of 1945. I learned that Dr. Rock was much more than a Harvard Medical School professor and fertility specialist. He also had worked passionately as the leading advocate on the legalities of fertility medicine to drive changes in policies.

During 1944, Dr. Rock had organized nearly one hundred fertility specialists nationwide to found the American Society for the Study of Fertility and Sterility. Organizations have clout. That was the reason he traveled to Chicago for a February 26, 1945, meeting with the nation's legal community at their annual symposium. The American Society for the Study of Fertility and Sterility successfully pressured the legal community to place the topic of artificial insemination on their agenda for discussion. Just a month before my conception, most likely at Dr. Rock's 10 Beacon Street office at the Free Hospital for Women in Brookline, Massachusetts, the 1945 Illinois Court's *Hoch v. Hoch* ruling had proclaimed donor-insemination as adulterous and the resulting child illegitimate.

Adding drama and confusion, just a few years later, the 1948 New York Superior Court ruled just the opposite. In a custody case, *Strnad v. Strnad,* the court granted a divorced father visitation rights to his child conceived via artificial insemination by donor, thus declaring the child legitimately "semi-adopted." Pope Pius XII further condemned the practice of artificial insemination by donor while presiding over the well-publicized 4th International Congress of Catholic Doctors in 1949. I wondered whether the Catholic Dr. John Rock, who had advocated opposing views, had attended that conference.

In spite of the flurry of contradictions, artificial insemination by donor gained practitioners throughout the 1940s and into the early 1950s among the small group of fertility specialists. With further research, it appeared to me that the freezing of sperm had served as one of the significant catalysts for a surge in artificial insemination by donor and its greater social and legal acceptance—which took an additional two decades to evolve, by way of the 1930s European farm.

In doing this exciting new round of post-discovery research, I learned that E. J. Perry brought the artificial insemination dairy cooperative approach from Denmark to the United States in 1937. As a result, seven fledgling cooperatives had sprung up by 1939. As these dairy cattle cooperatives grew, they needed to extend the life of stored bull semen and find safe ways to distribute it, and their needs far exceeded available technology. Recognizing the void, in 1939, two University of Wisconsin researchers, Dr. Paul Phillips and Dr. Henry Lardy, discovered that egg yolk, which is rich in phospholipids and lipoproteins, protected the sperm from temperature shock.

In 1941, Dr. Craig Salisbury, the son of an Ohio farmer, who had moved from his position as the head professor of animal husbandry at Cornell University to become the department head of the renamed and reorganized Department of Animal Sciences at the University of Illinois, built upon that Phillips and Lardy scientific egg yolk discovery. When he and his research team combined the egg yolk–semen mixture with sodium citrate and bicarbonate buffers, they protected sperm cells from temperature shock, which allowed for their cold storage, and extended the life of the semen for several days. That extended life enabled farmers to ship bull semen to fields some distance away. As a result, by 1949, dairy cattle conception rates vaulted by 15 percent.

By 1949, Dr. Christopher Polge, the son of an English poultry farmer, broke open the field of artificial insemination by donor while a researcher at the Division of Experimental Biology at the National Institute for Medical Research in London, and later at the Animal Research Station in Cambridge, England. His innovation, cryopreservation (the freezing of sperm), recorded its initial success (the first chicks bred via frozen sperm) by discovering that the combination of glycerol, dry ice, and alcohol protected the potency of chicken sperm from freezing damage. By late 1949, Harvard's Dr. John Rock documented that frozen human sperm remained fertile for a year. Yes, it was alive, but it remained in his laboratory, and was not used to create a pregnancy.

Also building on Polge's discovery, in 1950 Dr. Salisbury's former colleagues at Cornell University's Department of Animal Husbandry had innovated what the industry has termed "the Cornell extenders." They added antibiotics (penicillin, streptomycin, and polymyxin B) to the glycerol mix, which was cooled to 5 degrees centigrade to enable a longer life and a longer possible commute to its ultimate destination.

Dr. Polge built on that discovery. In 1953, he and his team enabled the birth of the first calf (cleverly named Frosty) conceived from his concoction of frozen bull semen. Polge improved his process and later replaced dry ice with liquid nitrogen, which yielded much colder, longer, and safer storage conditions. Dr. Polge's process immediately catapulted to prominence within the dairy community. By the mid-1950s, nearly 75 percent of dairy breeders, some from as far away as Argentina, were artificially inseminating their dairy cows with the sperm from champion bulls.

An inevitable consolidation also occurred in the dairy cooperatives to form the handful of large companies that exist today. The two largest companies in the dairy field serve as both dairy cooperatives and global award-winning bull sperm distributors. ABS Global of DeForest, Wisconsin, sells nearly 3.5 million breeding services per year. Not far behind is Select Sires of Plain City, Ohio. While 60 percent of US dairy cows are artificially inseminated, the European dairy industry's artificial insemination conception rate, led by Denmark, Holland, and England, exceeds 90 percent.

Frosty's 1953 birth foreshadowed the breakthrough in the freezing and thawing of human sperm at another university noted for its research in livestock. In the University of Iowa's newly formed fertility clinic, graduate student researchers Raymond Bunge, an ambitious urologist, and Jerome Sherman, who had experimented on the freezing and thawing of his own sperm, released their groundbreaking research results on the preservation of human sperm. They found that human sperm (which is far more fragile than bull sperm) remained fertile when combined with glycerol (which is used in low-fat baking), slow cooled, placed in refrigerated storage using carbon dioxide, and thawed gradually. By the time their findings were published in the *Proceedings of the Society for Experimental Biology and Medicine*, the doctors had artificially inseminated and impregnated three women with frozen sperm, an all-time first.

Sherman and Bunge wrote up their results for the prestigious American journal *Science*, but were denied for publication until they had proof that healthy babies, not faulty sperm-compromised babies, would be born as a result of the conceptions. When the women were three months pregnant, the doctors provided X-rays of their fetal skeletons, which showed that their development was progressing normally. The scientists submitted their paper to the British journal *Nature*, where it was published in October 1953 along with a press release titled "Women Pregnant by Frozen Human Sperm." The *New York Times*, among others, latched onto the news with an article titled "Baby from Frozen Sperm Expected in Three Months."

By early 1954, it appeared that three healthy babies had been born from that frozen sperm, until one child began having seizures and was diagnosed as blind in one eye. It took several months to determine that the pregnant mother's exposure to toxoplasmosis, a parasite from a cat's litter box, and not a flaw in the frozen sperm, was the culprit.

Meanwhile, in Iowa, the hate mail poured in. After being labeled "a scientific monster," "unchristian," and "a disgrace to medicine," Dr. Bunge backed away from the limelight, dropped out of the reproductive science field altogether, and returned to the less-heated and uncontroversial kitchen of urology. Dr. Sherman took a position at the University of Arkansas to continue his advocacy for cryobiology and assisted reproductive science. And society's support slowly inched in his direction. In a nationwide poll, 28 percent of Americans approved of artificial insemination. Donor-insemination had emerged from its relatively low profile to become a worldwide phenomenon. Also in 1954, *The British Medical Journal* published the first comprehensive methodology of the process of artificial insemination by donor.

While artificial insemination by donor was no secret in the 1950s, Sherman and Bunge's first success with frozen human sperm began to pave the way for greater social acceptance of artificial insemination by donor. Over the next two decades, the atmosphere practicing fertility physicians and their patients were in was peppered with continuing Roman Catholic Church condemnation, conflicting legal rulings on the legitimacy of a donor-conceived child, and various articles, both sympathetic and critical, in the popular press.

On the legal front, in 1956, the Illinois Superior Court ruled in *Doornbos v. Doornbos* that a child conceived via artificial insemination by donor was indeed illegitimate, but that a consenting husband was held liable for that child's support. Conflicting legal rulings were handed down across the globe. A 1956 court in Rome, headquarters of the Roman Catholic Church, labeled artificial insemination by donor as adulterous. A 1958 Scottish court ruled that a child from artificial insemination by donor was considered protected under the laws of adoption. A 1963 Superior Court in New York legitimized children of the "semi-adoption" process of artificial insemination by donor. A California court followed suit.

Given the increased cultural awareness and tilt toward acceptance of artificial insemination by donor throughout the 1950s and 1960s, fertility as a medical specialty had increased its global footprint and had an increasing number of practitioners. The code of secrecy according to

Seymour and Koerner's unofficial handbook remained the prevailing practice, though there were some harsh critics of the practice, including popular physicians such as Dr. Alan F. Guttmacher, an obstetrician and gynecologist at Johns Hopkins Hospital. His writings for both medical and lay audiences were published from the 1930s until his death in 1974. He had boldly challenged the Seymour and Koerner claims of donor-conceived births, survey findings, and pieces of their approach. But his methodology did not appear that much different to me.

While Guttmacher had also adhered to the code of secrecy, he argued that fertility practitioners should lower their patient fees and also serve as delivering birth physicians, since, in his view, the bond of trust developed between doctors and patients could best represent the white lie of paternity on the baby's birth certificate. His approach also included a resurgence of eugenics; he served only couples whom he deemed acceptable, solicited and purchased sperm from his medical students, screened donors' heredity, and tested them for venereal diseases. Several factors were important in his sperm donor selection process: their physical characteristics, religion, and ethnicity.

In 1952, Guttmacher moved to New York City as director of obstetrics and gynecology at Mount Sinai Hospital, where he flourished throughout the 1960s as a high-profile family planning advocate and staunch ally to Dr. Rock and Dr. Pincus (who co-created the birth control pill), and was a frequent congressional lobbyist.

In the midst of the introduction of the revolutionary birth control pill in 1961, Italian medical researcher Daniele Petrucci announced that he had succeeded in culturing a human embryo for twenty-one days. The Roman Catholic Church had taken a strong stance on the matter and viewed all reproductive science as an enemy to church doctrine. The Roman Catholic Church quickly condemned both Rock's pill and Petrucci's research. The evolution of sperm banks added to their strategic conflict.

The concept of a sperm bank and the practical use of frozen sperm in artificial insemination had not been considered until the early 1950s. It took several decades for the concept of sperm banks, plus the freezing of sperm, eggs, and embryos, to take hold. It required over three decades of refinement and sociological change to emerge as a commercially viable enterprise. Social commentators in the 1960s seized upon the *Cedar Rapids Gazette*'s 1953 feature of frozen fatherhood after death or disablement and added *Brave New World* commentary about parents selecting their ideal qualities (character, intelligence, aptitude, race, appearance) of a child's biological father from a well-stocked catalog of desired male specimens.

I discovered that the frozen sperm bank's first spawning had not targeted those requiring conception by artificial insemination by donor, but rather those who had sought to preserve their own ability to reproduce to thwart their potential infertility. Insiders coined it "insurance sperm." The combination of the evolution of the post-WWII vasectomy as a method of male birth control and the sterility-inducing chemotherapy treatment for cancer had both provided the catalysts for the commercialization of the frozen sperm bank. Early customers who faced those threats to their fertility had their own sperm frozen as insurance.

Throughout the 1960s and 1970s, the advocates of positive eugenics latched on to the sperm bank concept to enable a "superior stock" of healthier children. In 1965, Nobel Prize–winning geneticist Hermann Muller advocated for a "seminal Fort Knox" of those with idealized physical, mental, and personality characteristics. Fearing a nuclear holocaust and having discovered the inherited impact of radiation poisoning upon human chromosomes, his vision of a selective storage bin would preclude reproduction by those with less than ideal genetics.

Following Muller's commentary, *Science Digest* featured the respected director of the Center for Research and Training in Reproductive Biology at the University of Michigan, Dr. S. Jan Behrman, on the topic of the eugenics potential of frozen sperm. He had prognosticated on the futuristic likelihood of isolating chromosomes to recreate the persona of another Einstein or Beethoven.

"Insurance" sperm wasn't a large enough market to enable a thriving, profitable frozen sperm industry. The assisted reproductive technology field (the fertility physicians) were their targeted distribution channels of choice. Even though sperm bank operators had promoted statistics showing that children conceived via thawed sperm were healthier, had fewer birth defects, and were of "superior stock" (the eugenicists at work), fertility specialists had seen that fresh sperm achieved a meaningfully higher conception success rate. Therefore, they continued their practice of soliciting sperm donors on an as-needed basis, often seeking donors in and around their medical schools, and generally artificially inseminated their patients with a donor's fresh sperm. Faced with logistical issues regarding the timing and availability of a donor, a small but active subset of doctors took a shortcut and unethically used their own sperm . . . anonymously.

Artificial insemination by donor via an underground cadre of specialty physicians who employed anonymous ghost fathers and a code of secrecy to prevent the resulting children of couples with infertile husbands from being labeled "artificial bastards" was the fertility practice standard for much of the twentieth century. Beginning around the time that I passed into adulthood as a college graduate in 1967, society's rules began to change.

Within a milieu of societal upheaval, fertility science, a global health epidemic, and unrelenting judicial lobbying, the market drivers of the fertility industry began to change. Frozen sperm from a bank was legitimized as an option, and the sociological acceptance of the practice of artificial insemination by donor and its byproduct, donor-conceived people like me, accelerated. Surveys claimed that donor-conceived births in the United States were expected to grow by 6,000 to 10,000 per year, and that patients were served by 500 fertility specialists. In addition to treating infertility, 26 percent of those 500 physicians incorporated artificial insemination to prevent the transmission of genetic diseases. In other words, an unhealthy husband could opt to use the sperm of a healthy donor to increase the likelihood that he could be a dad to a healthy child.

Because physicians continued to prefer fresh sperm over frozen sperm for artificial insemination, the fledgling sperm bank industry had to find ways to improve their market position throughout the 1970s. They developed a "cryofreezer" that froze pellets of sperm in twenty minutes; they used liquid nitrogen to freeze sperm to –196 degrees centigrade to preserve sperm for up to ten years; they promoted access to the same donor for multiple inseminations; they cataloged their donors in eugenics fashion and marketed the imagery of their "desirable" characteristics in the style of a Proctor & Gamble advertising specialist as opposed to a medical practitioner; and they promoted the logistical ease of catalog shopping for the "right" donor versus the process of doctors recruiting and screening live donors on an as-needed basis. Still, by the end of the decade, only approximately 1,000 babies out of an estimated 100,000 babies conceived via artificial insemination by donor worldwide were conceived from frozen sperm. *Fertility and Sterility* noted in a 1978 article that "the early enthusiasm for using frozen semen has been tempered. . . . The ideal method for freezing gametes has not yet been found, and the commercialization of sperm banking has not developed. . . ."[15]

In 1973, the year my daughter was born, an unsubstantiated claim was made by Douglas Bevis, a noted British gynecologist, that he had achieved three IVF births. By the summer of 1978, while the frozen sperm field was struggling to gain its foothold, I was busy living my life. While my family celebrated Tracy's fifth birthday with chocolate cake

15   Edward Wallach and Rudi Ansbacher, "Modern Trends: Artificial Insemination with Frozen Spermatozoa," *Fertility and Sterility*, Volume 29, Issue 4 (April 1978): 375–379.

in my mother's backyard during a Cape Cod vacation, reproductive science took a giant technological leap. Susan and I were vaguely aware of this advance, but I would bet that my mother was fully informed of it. She never discussed it—another opportunity to come clean squandered. The birth of Louise Brown in a hospital near Cambridge University gained worldwide acclaim, the very first "test tube baby" by IVF, thanks to a decade of work by a trio of English researchers. Harvard's Dr. John Rock had succeeded in fertilizing a human egg with sperm in a laboratory in 1944. But in Great Britain, Dr. Patrick Steptoe, Dr. Robert Edwards, and nurse Jean Purdy succeeded in impregnating their female patient, who had blocked and distorted fallopian tubes, by laparoscopically extracting her egg, fertilizing it in their lab with her husband's sperm, and planting that resulting embryo into her uterus for a full-term pregnancy and the birth of a perfectly normal, healthy baby girl—artificial insemination by husband via IVF. By 1980, they had refined and documented their process another sixty-eight times and went on to win the Nobel Prize. A cadre of fertility specialists around the world took note and expanded their practices several fold as a result. The first IVFs were quickly recorded in India, Australia, and the United States. By 1982, doctors in France, Sweden, and Austria recorded their countries' first IVF births, also using the laparoscopic technique perfected by Dr. Steptoe.

Edwin Carlyle (Carl) Wood of Monash University in Australia leapfrogged reproductive science when he announced two of his own firsts in 1983: the first IVF baby from a frozen embryo, followed quickly by the first egg-donor baby born to a menopausal woman.

In 1984, Copenhagen-based researcher Dr. Susan Lenz, in concert with Dr. Wilfried Feichtinger of Vienna, developed a new standard: the first use of a new technique, the far less traumatic ultrasound-guided transvaginal egg retrieval under local anesthesia. She retrieved the eggs via a needle inserted through the bladder. The University of California, Los Angeles (UCLA), quickly followed suit with an egg donor birth that used the ultrasound retrieval technique. Owen, by then my eighty-year-old former mentor, commented, "Leave it to research scientists at UCLA to use the 1984 Orwellian date." By 1985, a trio of Swedish practitioners, Matts Wikland, Lars Hamberger, and Lars Nilsson, further refined the use of the ultrasound-guided transvaginal technique for egg retrieval.

First, anonymous sperm donors, now anonymous egg donors, I thought, *creating another group of people like me to encounter the same genealogical bewilderment.* Fertility specialists immediately adopted the tried-and-true Seymour and Koerner handbook with a eugenics tone for the selection of egg donors, with secrecy and anonymity at its core.

Dr. John Charles Rock fully retired in 1976 and died in 1984. By 1995 (when I learned that I was donor-conceived), his clinic and the hospital itself had disappeared, both in name and from the physical landscape. While his office space disappeared (the old hospital sold to developers in 1989 and converted to luxury condominiums), his legacy was permanently etched in the history books. As I researched, I felt that the man was under-recognized by society as a whole for his profound influence on reproductive medicine. Dr. Rock's American Society for the Study of Fertility and Sterility was renamed the American Fertility Society and then the American Society for Reproductive Medicine (ASRM). While "American" in name, it is active in one hundred countries. Fertility became one segment in the field of reproductive science.

Just a year after Dr. Rock's death came the world's first human "gestational surrogacy." A fertilized embryo from a couple had been implanted (for compensation) into another woman's womb (the surrogate) to yield a full-term, healthy baby.

But the 1986 "Baby M" case obscured those 1985 headline achievements. A surrogate mother refused to relinquish custody of the baby. The New Jersey Supreme Court ultimately ruled that the intended parents have custody, but the surrogate mother has visitation rights.

The Baby M headlines buried another first in the 1986 flurry of reproductive discoveries. Singapore's Dr. Christopher Chen reported the world's first pregnancy using a frozen egg. By freezing their eggs, women facing sterilization from cancer-treating chemotherapy or radiation, seeking to delay pregnancy to build their careers, or without suitable partners at the moment could preserve their ability to conceive a child.

Six hundred surrogate babies were born within the first five years of surrogacy's 1985 debut. No longer operating in the shadows, reproductive scientists had become Nobel Prize–winning global superstars in their own right. The Roman Catholic Church swallowed hard, just as it did when the Rock/Pincus birth control pill captured worldwide acclaim in the 1960s, but by this time large segments of multiple societies had by and large stopped heeding Roman Catholic Church doctrine. Fertility as a medical practice flourished openly. A 1988 Congressional Office of Technology survey reported that 11,000 fertility physicians served in clinics in the United States.

When it came to sperm donation to create a pregnancy, however, global fertility practitioners mostly continued to acquire fresh sperm from medical students, residents, or physician donors instead of using frozen sperm. Just as fertility practitioners had found a way to market their multiple services to a larger group of people, so, too, had the purveyors of frozen sperm prepared to make the same leap in growth.

The market driver? The late 1980s AIDS epidemic.

Biological researchers discovered that bodily fluids (sperm and blood) were HIV carriers. With several AIDS infections reported from freshly donated sperm in the United States, Canada, and Australia, sperm banks went into overdrive as a fertility doctor's safe choice. Freezing allowed for the quarantining and testing of the sperm for HIV and other infectious diseases.

The sperm bank industry grew in Wild West fashion, from ten

sperm banks in 1969 to 135 in the United States in 1990 which were, by all accounts, loosely regulated, unaccredited, and disorganized. By 1995, with no professional association representing the various sperm banks, the American Association of Tissue Banks (AATB) stepped in to offer accreditation for a fee. Just seven of the 135 signed up and gained that accreditation.

A number of those sperm banks were mom-and-pop operations, sometimes run out of an individual physician's office. The increased costs of doing business—to recruit healthy donors, screen for AIDS and other diseases, and store, package, market, and distribute the "product" led to inevitable consolidation to create a few larger enterprises that could afford to indefinitely store and market thousands of specimens from one location.

By the end of 2001, with the internet bubble spectacularly burst and the 9/11-induced recession underway, twenty-eight fully functional sperm banks existed in the United States, according to the AATB and a newly published Sperm Bank Directory. Concentrated in the East Coast, West, and upper Midwest, they served fertility specialists from their locations in sixteen different states. And they fueled the reproduction of multiple siblings from the same sperm donor. The largest five of those banks, with approximately one hundred donors each, supplied almost half of the donor sperm required nationally.

My friend Mike and I dissected these statistics together. From data spanning nearly three decades, we could estimate that three donor-siblings were born each year. If we used the old 80–20 rule, where 80 percent of the recipients might prefer and select 20 percent of the donor pool, the numbers of donor-siblings could well have surpassed twelve annually. With a ten-year life, is it any wonder why donated frozen sperm has accounted for one hundred or more offspring from the same donor? We also looked at the business math as sperm banks actively recruited donors on college campuses. To bypass regulations on selling human tissue, donations are labeled a "service." Fees cover "time and travel." At $100 per donation, often three times per week for a month, a donor could earn $1,200 in cash, with no 1099 tax accountability—under-the-table money he could keep. Some donated for a year or more, earning over $14,000 per year tax free. The sperm bank marked that cost up by as much as 4,000 percent; enviable gross margins, to be sure. I found that the closest thing to a regulatory authority was Dr. Rock's American Society for Reproductive Medicine, which had updated the Seymour and Koerner "handbook" of "best practice" recommendations.

The ASRM's practice guideline limits gamete uses per donor to twenty-five per population of 800,000. Is that a limit? Using that guidance, I would have two dozen siblings in Sacramento, over one hundred siblings in metropolitan Boston, and two hundred fifty siblings in New York City or Los Angeles. Other countries had enacted laws that put clear caps on the number of births per donor to lessen the chances of an inadvertent incestuous marriage of siblings. The United States had no such laws. These sperm banks operated like Harrods or Nordstrom. They had return policies, warranties, credit card acceptances, FedEx shipments, catalog models (according to traits, education, blood type, race, etc.), and internet marketing.

I discovered that some countries banned or restricted compensation for sperm donation. For instance, Sweden banned anonymous donations in 1980. While the UK did so in 2005, it provided no mechanism by which people who had previously been conceived with donated sperm could gain information about their donor. Canada outlawed paid sperm donation altogether. Collectively, such actions spawned "fertility tourism," where would-be customers, mostly women, traveled to foreign lands for artificial insemination by an anonymous donor. Alternatively, some nations exported their frozen sperm. Denmark led the pack, marketing their "desirable" Danish sperm (donors featured as altruistic, uncompensated, tall, fair-skinned, well-educated) to fifty countries.

As I concluded my sperm bank research, I came to realize it was a far-reaching global industry that would interest the likes of Morgan Stanley or Goldman Sachs; analysts were forecasting a $5 billion worldwide market by 2025, up from $3.5 billion in 2015. The FDA had banned the importation of sperm to the United States along the same lines as they had banned other potentially disease-bearing organic products, thus eliminating foreign competition.

During the expansion of the sperm bank industry, reproductive science leapt forward once more. In 1989, London-based reproductive endocrinologist Robert Winston and embryologist Alan Handyside reported the first successful identification by sex of three-day-old embryos. By 1993, the first successful birth using intracytoplasmic sperm injection (one sperm cell to fertilize one egg, also known as ICSI), moved reproductive technology forward rapidly. Since one ejaculation produces over 250 million sperm cells, these advances in science created the opportunity to produce multiple births from one donor's single deposit, and the expansion of artificial insemination by donor reached previously unimaginable astronomical proportions.

The first baby born using the mitochondrial DNA of a donor, the use of which created the so-called "three-parent embryo," took place in 1996. In the United States, the FDA, exercising its authority to regulate the genetic modification of organisms, halted the use of such techniques in the early twenty-first century.

Singapore-based researcher Dr. Lilia Kuleshova had perfected an egg freezing technique in 1999 that enabled a forty-seven-year-old post-menopausal woman to give birth to a healthy baby girl. Surrogate grandmothers were in the news, carrying and bearing children on behalf of their daughters. Egg donors, also actively recruited by the fertility trade, could earn many times more than sperm donors—$10,000 or more per donation—anonymously and without tax consequences. Because the process of donating eggs is far more complicated than donating sperm (donating eggs requires hormonal preparation and vaginal ultrasound-guided extraction while under sedation instead of simply masturbation), the compensation is far greater. A few months of egg donation could earn enough to buy a new car, pay off a college loan, or put a down payment on a house.

Meanwhile, fertility specialists had also discovered methods for empowering weak sperm, either by hormonal injections or gathering them in a laboratory and reinforcing them to ensure potency. With that reinforced sperm, formerly sub-fertile males, with the help of IVF, were impregnating their wives or partners.

The old definition of family was indelibly altered as society moved into the twenty-first century. From the time of my childhood until a decade before my California granddaughter was born in 2012, the percent of children born out of wedlock or to unwed parents in the United States moved from under 5 percent to over 40 percent. A marriage certificate was no longer the primary marker of a long-term, committed couple. Nor was heterosexuality. In 1996, Growing Generations, the first surrogacy agency to provide services to same-sex couples, opened in Los Angeles. Old laws were clearly antiquated, and the power of new marital practices finally overcame the resistance to change. The United States legal community lobbying that had been instigated by Dr. John Rock in 1944 at the Chicago Symposium was finally set into law in three different waves under the Uniform Parentage Act.

In its first wave, the 1973 Uniform Parentage Act (UPA) removed the label "illegitimate" from a child regardless of the marital status of the parents. In 2002, the second wave of the UPA made its way through the courts; sixty-eight years earlier, Dr. Rock would have celebrated joyfully with his colleagues. A donor-conceived child or surrogate-conceived child was deemed "legitimate." His or her *consenting* parents (defined as a man and a woman) were deemed the child's legal parents. The donor had no paternity right or obligation.

Time's "Artificial Bastard" and his adulterous mother had, at long last, disappeared! Gone, too, was the need for utmost secrecy and the practice of a birth certificate hoax that gave me my semi-adopted name. Whether to disclose information to children about their origins was a decision that remained with the parents, but the ingrained practice of secrecy and donor anonymity was modestly dislodged. The third wave of the UPA in 2015 reflected the legalization of same-sex marriage and granted those same donor-conceived rights to same-sex couples and their children.

The clients who sought out donated sperm reflect changing demographics: 20 percent were different-sex couples, 60 percent were same-sex couples, and 20 percent were single women. Although legal precedents have now been established throughout the Western world, many countries with more traditional sociocultural norms have yet to accept or approve artificial insemination by donor for any couple, different sex or same sex, or for single women.

Donor-conception (the old semi-adoption) and adoption have followed similar pathways regarding the rights of the child; both began by keeping parental information a secret. Closed adoptions began to change in the 1970s. Adoption and psychology professionals cited the reported destruction closed adoption could inflict on the emotional and developmental health of adopted children; genealogical bewilderment could spill over to negative self-esteem. They concluded that open adoption was far healthier for a child's psyche. Except in certain cases, for instance assault or severe abuse, open adoption has become the general practice since the 1990s. Open donor-conception practices fell severely behind adoption practices.

Statistics have not been formally updated since 2010, when the ASRM reported that one million people were donor-conceived and forecast an increase of 30,000 to 60,000 annually (five to six times greater than the 1970 estimate). Over the same year, 136,000 children of varying ages were adopted in the United States. Doing some quick math, if I use the new donor-conceived birth midpoint of that range, I estimate that almost 1.5 million donor-conceived individuals existed in 2020, an astounding increase of 50 percent over the past ten years. For the most part, those individuals were brought into this world like I was—semi-adopted in closed fashion via an anonymous donor, believing that their inherited biology was something that it was not.

The fertility physician's updated "best practices" playbook written by the ASRM highly recommends that parents provide open information to their donor-conceived children. Practices vary from sharing information very early to disclosing information at the age of eighteen, but it is a recommendation, not a law, in the United States (although it is law in some countries); the recommendation is subjectively implemented and a matter of individual parental choice. But their hand is often forced by the consumer DNA information that has become a commodity on the internet—the "oops" factor.

I learned two new terms during my research and my personal DNA experience: "misattributed" and "Non-Parental Event," or NPE. The NPE Fellowship (npefellowship.org) estimates that 2 to 4 percent of the world's population is "misattributed." That is a jaw-dropping number by itself. Some estimates are much higher. Late discovery of closed adoption; conception by anonymous sperm, egg, or embryo donation; or a surrogate carrier are just pieces of that puzzle. Added to those possible methods of conception are an extra-marital affair, a one-night stand, or sexual assault. And to add even more possibilities to that mix, one could have been switched at birth or raised as a child of another family member (an aunt, an uncle, a grandparent, or a sibling).

By 2020, DNA firms collectively had sampled the DNA of nearly fifty million customers, a low-end 2 percent estimate (or one million) of whom have experienced the unintended consequence, the "oops" factor. In geometric proportions it is actually greater than one million; 2 percent of previous generations (one's parents, grandparents, great-grandparents, and so on) could well be unknowingly misattributed. For example, using the 2 percent estimate, in my high school graduating class of one hundred students, two of us were directly misattributed to one or both of our parents (myself and another of my classmates), and twenty-eight of us were misattributed through our great-grandparents (exceeding one in four). Using that 2 percent factor and going back to our sixth great-grandparents would likely show that all one hundred of my classmates had at least one direct ancestor in their supposed family tree that is absent from their actual gene pool. Even without taking into consideration the myriad great aunts and uncles or distant cousins who may also be impacted by that 2 percent misattribution factor, all the family stories and family lore have been placed on the chopping block in millions of families across the world since DNA over the internet was innovated by 23andme and recognized in 2008 as *Time*'s invention of the year. Unanswered questions remain. DNA does not lie. So many family secrets are artificial. There is no place to run; no place to hide. Truth is in open view.

I was purposely misattributed in 1945 to my infertile dad on my birth certificate in an elaborate hoax that used an anonymous sperm donor. Had my mother taken that truth to her grave, as she had originally intended, and had I recreationally sought to analyze my DNA, the "oops" factor would have kicked in immediately. What I experienced was traumatic enough; imagine the disruption to my fragile and flawed psyche if I had not been prepared for the results!

That scenario has been the substance of countless newspaper headlines, magazine articles, and documentary films since the DNA analysis revolution took hold. The "oops" factor has captured its place in world culture. They started as good news stories like mine: twenty-two years of searching that ended with me finding my genetic paternal roots and gaining a cool sister to boot. Others told stories in which childhood best friends discovered they are biological brothers. Then the stories got ugly: One woman from Dr. Mary Barton's WWII–era London clinic had six hundred siblings (courtesy of Barton's sperm-donating husband). A medical student who donated to a clinic with the promise of no more than five inseminations discovered as an adult doctor that he has nineteen offspring (and still counting!). One doctor used his own sperm for artificial insemination dozens of times instead of using sperm from the patient's selected and registered donor from a sperm bank. A woman and her natural children were riddled with medical issues resulting from little or no donor screening. Incorrect sperm bank records resulted in racial or other characteristics that differed from those of the donor selected by the parents. A dozen-times donor professed to be a genius who spoke four languages and was, in fact, a convicted felon with schizophrenia. New stories pop up all too frequently that challenge my imagination.

The "oops" factor, when discovered, compounds the secrecy issues. There is no prosecution for outright fertility fraud; it is unethical, not illegal. As I discovered in my genetics on 23andme and later on Ancestry.com, I was keenly aware that I, too, might open myself up to the trauma of an "oops" discovery. How fortunate I was that my parents had trusted a class act, Dr. John Charles Rock. My biological roots are not from a horse thief, a disease-ridden fleabag, or an unethical fertility doctor who used his own sperm multiple times; rather, I come from an upstanding, healthy, intelligent guy.

During Dr. Rock's lifetime, he did his utmost to encourage the legal community to address the legal status of donor-conceived children. Given those efforts, if Dr. Rock had lived to see our scientific and sociological changes, would he have vigorously lobbied for legal regulations that would restrict the number of recipients of frozen, banked donor sperm (or egg/embryo) per donor lest any community unknowingly house multiple dozens of half-siblings? Would he have supported regulations and sanctions to outlaw and punish the unethical and repeated use by a fertility doctor of his own sperm as the anonymous sperm donor? Would he have led a change in fertility practice standards that also supported identity disclosure, not only for the offspring of sperm, egg, or embryo donors, but also for adoptive children? DNA analysis has made that old secret practice obsolete. And what would he have thought upon seeing advertisements for sperm, egg, and embryo donors or surrogate carriers on Craigslist and artificial insemination kits on Amazon.com? The answers are all speculative.

Not speculative are the exponential advances of science and society's embrace of the advances we have witnessed thus far. IVF using eggs as well as sperm, with the eggs carried to full term by surrogate mothers (this has become a major import/export industry in the Ukraine), has advanced reproductive capabilities. Those eggs may have originated from the child's intended mother or from another woman, an egg donor. We have already seen the first IVF babies born in England and the United States, the first successful egg donation at UCLA, the first baby born from a frozen embryo, and the first child born following a pre-implantation genetic diagnosis.

By 2014, the University of Gothenburg had upended reproductive science once again. A team led by doctor-professor Mats Brännström recorded the first baby born from a transplanted uterus. The thirty-six-year old woman who gave birth had been born without a uterus; she had received a transplanted uterus from a family friend, a sixty-one-year-old donor.

Where will advancing reproductive science take us?

As it has throughout history, society has continued to adapt to expediential scientific advances—sometimes kicking and screaming, but it has adapted, nonetheless. Humans have swept up innumerable pieces of emotions and lives shattered by the unintended consequences of disruptions that have challenged social conventions, all in the name of technological progress. That advancement has not traveled on a straight line, however.

On January 6, 2021, journalist Nellie Bowles's article in the *New York Times*, titled "The Sperm Kings Have a Problem: Too Much Demand," featured an internet underground encouraged by the coronavirus pandemic and nurtured by social media. The pandemic had created a problem: Some work-at-home women had more time to have a child, far fewer men were violating stay-at-home directives to make deposits at sperm banks, and demand for sperm was up 20 percent. The solution: competing social media groups offering direct-to-consumer sperm donations and smart phone applications to enable the match. Sperm Donation USA and USA Sperm Donation are private Facebook groups sporting tens of thousands of members. Similar groups exist in varying countries to offer "direct deposit, no charge: just pay for travel." To help make the right match (college educated, tall, slender, attractive, personable, desired skin tone, hair color, and eye color), software applications and websites have sprung up that look like Match.com. In online dating fashion, sites like Modamily, Known Donor Registry, and Just a Baby have attracted a database exceeding 50,000 members who compose the Etsy of sperm donation (customized by size, shape, color, and traits), with no required adherence to ASRM guidelines. An under-regulated sperm bank industry has new-age competition in the shape of an online flea market with no regulatory oversight. *Dr. Rock would turn over in his Marlborough, Massachusetts, grave*, I thought.

As the world turns, reproductive science has developed its capacity to select the sex, the healthy genes, and even the traits of the unborn child. Society continues to experience heart palpitations over this technology; our leaders fear that these new discoveries smack of eugenics and have outlawed their use. Precedent-setting Chinese reproductive researcher Dr. He Jiankui earned the dubious reward of a three-year jail sentence in 2019 for his pioneering gene editing of twin girls in his attempt to prevent transmission of the HIV virus. What nickname might history give him?

And how might cloning technology, first applied to farm animals (sound familiar?), challenge society's sense of morality when it is applied to humans via the next Red Frankenstein? In 2021, US scientists announced the first cloning of an endangered species, a black-footed ferret, from the DNA retrieved from the frozen remains of a ferret that had died in 1988. Can one imagine a cloned Einstein or Beethoven? How about a frightening monster: another Hitler or Manson?

The Bill and Melinda Gates Foundation has vociferously advocated for CRISPR technology in its movement to reestablish the ground rules in support of gene editing to benefit humanity. Will that once-discredited Chinese doctor's status move from ex-con to courageous, visionary hero—the Go-to Guy in his field? According to the scientists, CRISPR is a specialized region of DNA that creates what amount to genetic scissors—enzymes that allow the cell (or a scientist) to precisely edit other DNA or its sister molecule, RNA. For instance, as a virus attacks the body, the bacterium stores small chunks within its own DNA. This helps the bacterium recognize viral infections when they occur again. CRISPR-associated enzymes disarm the virus and prevent the infection from spreading. Interestingly enough, the use of RNA sits at the cornerstone of the 2020 successful and highly effective COVID-19 vaccine development of both Moderna and Pfizer.

It has been speculated that the Gates Foundation may invest tens of millions of dollars in gene editing that targets malaria and other mosquito-borne diseases to change the insects' genome to prevent them from passing along the parasites that cause those diseases. The foundation may put millions into water, weather, and the social services required to support a growing and aging population that is healthier than ever and that will be competing for the resources to enable it to flourish.

Science fiction writers have used powerful new discoveries to build their futuristic sagas. Reproductive scientist Shanna Swan, along with co-author Stacey Colino, in their 2021 book *Count Down*, warned of a global fertility crisis. Western-world sperm counts have plunged more than 50 percent over the past four decades because of impacts from toxic environments and an aging population that delays having children until after their peak fertility capacity. The trend line through 2045 threatens the survival of the human race. Combine this finding with the 2021 reproductive breakthrough in which model embryos were created from skin cells, without the benefit of uniting an egg cell with a sperm cell (published in the science journal *Nature*, by a Monash University team led by Professor Jose Polo), and creative imaginations are free to run wild.

Our global society and its economy continue to experience the inevitable bumps and hiccups as they change. Yet to come, for instance, is the move from fossil fuels to clean energy as the new technology inevitably progresses and we experience hurricane-strength pressure to adopt the clean technology. Other bumps are impacting DNA science. Growth in the entire consumer DNA sector slowed not long after it revealed my paternal identity—not only my genetics, but my ghost father as well, and a new close friend, a sister to boot! That growth spiked once again as people sought outlets for their boredom during the 2020 COVID-19 pandemic lockdowns.

The DNA-over-the-internet vendors have unbundled their consumer offerings in an attempt to increase their revenue. Health impact, digitized family records, and other services have become costly add-ons to basic ethnic analysis. What was initially a relatively low-priced consumer product has become a much higher priced, multi-product offering. The company noted for digitized records used to build family trees moved into the DNA health impact sector. The one noted for DNA health impact moved into family trees. Their differentiation is in danger of becoming commoditized and rather unrecognizable. The

Early Adopters are saturating the DNA analysis market. The previous prospect of rapid growth has tanked as the remaining population of would-be customers are finding diminished value in determining that they are 17 percent Irish, 22 percent Italian, 36 percent Persian, or whatever. Privacy concerns and the use of DNA analysis by the FBI, the CIA, and worldwide crime solvers, as well as sinister forces, all are dissuading new prospects and are containing its initial spiking growth. It is too early to see the impacts of the strategic moves by Ancestry.com and 23andme, though they have the backing of their new shareholders and are increasingly scrutinized by the public.

But all of this is not dissuading the curiosity of the donor-conceived or the adopted. In studies published in the journal *Human Reproduction*, 92 percent of the one million donor-conceived people over eighteen years old were searching for their donor, siblings, or both; 82 percent are so curious about their donor's physical appearance that they hope to one day meet. I would have searched feverishly for the rest of my days. I was not going to be denied. I would speculate that the five million adoptees in the United States who were adopted in closed adoptions since 1945, when I was born, would feel the same way once they learned that they were adopted. Let's not rule out the over 150 million people, 2 percent of the 7.8 billion worldwide population, who are potentially "misattributed" to one or both parents and likely do not know it—yet.

Meanwhile, I am finding an occasional sibling now and then. Oops!

The Donor-Conceived
Bill of Rights

Definitive human reproductive industry legislation will take the "oops" out of consideration, eliminate unnecessary trauma, and lend legitimacy to the practice of donor-insemination. From a donor-conceived viewpoint, this multi-billion-dollar human reproductive industry desperately needs to fix the gaps in its practices. A donor-conceived Bill of Rights would include the following requirements:

- Abolishment of donor anonymity

- Full and early disclosure of donor genetics
  and medical history

- Requirements for the genetic testing of donors

- Limitations on the number of offspring per donor

- Registered identification of sibling donor-offspring
  to one another

- Up-front counseling to donors and recipients regarding
  donor/recipient rights and responsibilities and the needs
  and rights of the donor-conceived child

- Defined legal consequences for blatant fertility fraud

Acknowledgments

I am enormously grateful to all who contributed to my discovery of the truth about my genetic origins and to those who advocated for and assisted with the process that led to me sharing this very personal odyssey and extensive research.

My family unselfishly rendered me their love, support, and encouragement throughout my emotional, topsy-turvy journey of search, discovery, and renewal . . . for a quarter-century. Susan, Stephen, and Tracy, my love is absolute.

My daughter, Tracy (the artist), has shown herself to be an incredibly resilient and creative forensic researcher extraordinaire. She could well begin another lucrative new career with her proven talents. Without Tracy's willingness to put those talents to work on her father's behalf, my genetic dilemma might never have been solved. I might never have grown to the degree that I feel that I have. She is the best!

My special appreciation goes to my "new" sister, Roxy, for her welcoming embrace. By adding the few key missing pieces needed to complete my genetic puzzle, Roxy has shown that she is an able sleuth in her own right. Is that the power of common DNA?

My friendship circle encouraged me to write a second book from the personal lens of a donor-conceived person once I shared with them my early research on the evolution of the secretive and sometimes scandalous practice of artificial insemination by donor. I fed off their interest and enthusiasm for what seemed to them to be a timely and informative potential book. Their supportive energy gave me the impetus to write an early draft. We collectively felt that the draft seemed incomplete without further research on the scientific advances in assisted reproductive technology that had occurred since my birth. The evolution of the legal and sociological environment had substantially eliminated a piece of the former secrecy (not all) from its methodology and altered the playing field.

I would like to recognize the unnamed and helpful staff of both the Boston Public Library and the Harvard Medical School's Countway Library and Center for the History of Medicine, who unflinchingly offered their time and talent to better direct my research and interpretation of early and more current assisted reproductive technology history.

Thank you to Barbara Monteiro, the New York City–based publicist who launched my first book, *All Hands on Deck: Navigating Your Team Through Crises, Getting Your Organization Unstuck, and Emerging Victorious.* She served as mentor, strategic consultant, cheerleader, introductory-agent, and friend once I shared my early manuscript with her. Barbara promoted this book to her network, which led to strategic help with shaping my story.

That good fortune led to my introduction to and development of a bond of trust with strategic editor Francine LaSala. Part therapist and part writing coach, she provided an empathetic influence that I found invaluable and enabled me to risk sharing personal intimacies—a risk that I might have resisted otherwise.

I offer a round of applause to the professional team at Greenleaf Book Group. They have taken my saga and turned it into a timely and helpful book that, I hope, will influence the practice of assisted reproductive technology positively for all participants (past, present, and future). My goal is that others will also find this book useful: the scientists, the reproductive practitioners, the donors, the gamete distributors (sperm, egg, and embryo), the users, the family lawyers and family therapists, the legislators, and those who are donor-conceived (like me) or misattributed in some other way.

I owe a special thank you to amazing IVF clinician Dr. David Ryley and highly accomplished life science executive Don Hardison. They each graciously opened their network to me.

Erin Jackson, founder of We Are Donor Conceived, allowed me to share my special circumstances with a unique group of people from around the globe who understand the personal trauma of learning later in life that they are not quite who they thought they were. I profoundly benefitted from their sharing the experiences that we all had in common: dealing with a genetic identity that had once been kept secret. I appreciate the openness of all the misattributed people and their willingness to assist others as they navigate the new knowledge of their unique heritage.

There is nothing like sharing and learning from others who have walked the walk.

Bibliography

1. Axelsen, Diana. "Women as Victims of Medical Experimentation: J Marion Sims' Surgery on Slave Women 1845–1850." Bioethics Research Laboratory, Georgetown University, 1985, Fall 2 (2) 10–13.

2. Berger, Joseph. "John Rock, Developer of the Pill and Authority on Fertility, Dies." *New York Times*, December 5, 1984.

3. Blyth, Eric, Marilyn Crawshaw, Lucy Frith, and Caroline Jones. "Donor-Conceived People's Views and Experiences of their Genetic Origins: A Critical Analysis of the Research Evidence." *Journal of Law and Medicine*, June 2012. https://www.researchgate.net/profile/Marilyn_Crawshaw/publication/230712679_Donor-conceived_People%27s_Views_and_Experiences_of_their_Genetic_Origins_A_Critical_Analysis_of_the_Research_Evidence/links/5764252d08ae1658e2ede3b7/Donor-conceived-Peoples-Views-and-Experiences-of-their-Genetic-Origins-A-Critical-Analysis-of-the-Research-Evidence.pdf?origin=publication_detail.

4. Bourn, Chris. "The Seedy History of Sperm Donation." MEL Magazine.com, 2017. https://melmagazine.com/en-us/story/the-seedy-history-of-sperm-donation.

5. Caldwell, John Harvey. "Babies by Scientific Selection." *Scientific American*, Volume 150, (1934).

6. The Center for Bioethics and Culture Network. http://www.cbc-network.org/.

7. The Center for Bioethics and Culture Network. "Three Things You Should Know about Sperm 'Donation.'" http://www.cbc-network.org/pdfs/3_Things_You_Should_Know_About_Sperm_Donation-Center_for_Bioethics_and_Culture.pdf?fbclid=IwAR2pbBddDVATJV0XQmVjDyZoM4jzfrAYdfJzlcKl9QF-lvoucecDsW2Uwx8.

8. Convention on the Rights of the Child, United Nations Human Rights Office of the High Commissioner. https://www.ohchr.org/en/professionalinterest/pages/crc.aspx.

9. Daniels, Cynthia R., and Janet Golden. "Procreative compounds: popular eugenics, artificial insemination and the rise of the American sperm banking industry." *Journal of Social History* (September 2004). https://www.thefreelibrary.com/Procreative+compounds%3A+popular+eugenics%2C+artificial+insemination+and...-a0123163927.

10. Davis, Frank P. *Impotency, Sterility, and Artificial Impregnation.* London: Henry Kimpton, 1917.

11. Fackerell, Michael. "Breaking the Bastard Curse." Christian Faith International Ministries, 2004.

12. "Family Law: Court Determines Child Conceived by Artificial Insemination to Be Illegitimate." *Duke Law Journal*, 1964. https://scholarship.law.duke.edu/cgi/viewcontent.cgi?article=1901&context=dlj.

13. "Ghost Fathers: Fathers for the Childless." *Newsweek*, May 1934.

14. Guttmacher, Alan F., John O. Haman, and John Macleod. "The Use of Donors for Artificial Insemination: A Survey of Current Practices." A Report of a Committee of the American Society for the Study of Sterility, Volume 1, Number 3 (1950). https://www.fertstert.org/article/S0015-0282(16)30188-1/pdf.

15. Haller, Dorothy L. "Bastardry and Farming in Victorian England." Loyola University Department of History, 1989–1990. http://people.loyno.edu/~history/journal/1989-0/haller.htm.

16. "The History of Sperm Donation." DonorUnknown.com. http://www.donorunknown.com/history.

17. Hix, Laura. "Modern Eugenics: Building a Better Person?" *Helix Magazine*, July 23, 2009. https://helix.northwestern.edu/article/modern-eugenics-building-better-person.

18. John C. Rock personal and professional papers, 1921–1985. Francis A. Countway Library of Medicine, Center for the History of Medicine. Boston: Harvard Library. https://hollisarchives.lib.harvard.edu/repositories/14/resources/4603.

19. Kramer, Wendy. "A Brief History of Donor Conception." HuffPost.com, May 10, 2016. https://www.huffpost.com/entry/a-brief-history-of-donor-conception_b_9814184.

20. Lynch, Sarah. "Fact Check: Father of Modern Gynecology Performed Experiments on Enslaved Black Women." *USA Today,* June 19, 2020.

21. MacGregor, Deborah Kuhn. *From Midwives to Medicine: The Birth of American Gynecology.* New Brunswick: Rutgers University Press, 1998.

22. Marsh, Margaret, and Wanda Ronner. *The Fertility Doctor: John Rock and the Reproductive Revolution.* Baltimore: Johns Hopkins University Press, 2008.

23. Marsh, Margaret, and Wanda Ronner. "Margaret Marsh and Wanda Ronner: Assisted Reproductive Technology and the Pursuit of Parenthood." Rutgers University, April, 2021. https://marshronner.rutgers.edu.

24. Marsh, Margaret, and Wanda Ronner. *The Pursuit of Parenthood: Reproductive Technology from Test-Tube Babies to Uterus Transplants.* Baltimore: Johns Hopkins University Press, 2019.

25. Massey, Albert P., Jr. "Artificial Insemination: The Law's Illegitimate Child." 9 VIII. L. Rev. 77, 1963. Villanova. https://digitalcommons.law.villanova.edu/cgi/viewcontent.cgi?article=1595&context=vlr.

26. McDougal, Sarah. "The Strange History of the 'Bastard' in Medieval Europe." Originally published at Aeon and republished under a Creative Common License on *The Wire,* June 18, 2017. https://thewire.in/history/history-bastard-europe.

27. "Medicine: Artificial Bastards?" *Time* Magazine, February 26, 1945. http://content.time.com/time/subscriber/article/0,33009,792012,00.html.

28. Moore, Geoffrey. *Crossing the Chasm.* New York: Harper Collins, 1991.

29. Ombelet, W., and J. Van Robays. "Artificial Insemination: History Hurdles and Milestones." Facts, views & vision in *ObGyn,* vol. 7, 2 (2015): 137–43. https://www.ncbi.nlm.nih.gov/pmc/articles/PMC4498171/.

30. Ravelingien, A., V. Provoost, and G. Pennings. "Donor-conceived children looking for their sperm donor: what do they want to know?" Facts, views & vision in *ObGyn,* vol. 5, 4 (2013): 257–64. https://www.ncbi.nlm.nih.gov/pmc/articles/PMC3987373/?fbclid=IwAR09zyTLMnZTXDZAlI-30UN3ycrxHALCfPgV4MZRSjaJsnVcrsH4O4XTyZ4.

31. "Roger Eugene Wilcox." Originally published in *The Arizona Republic,* August 2, 2006. https://www.legacy.com/obituaries/azcentral/obituary.aspx?n=roger-eugene-wilcox&pid=18724792.

32. Schellen, A. M. C. M. *Artificial Insemination in the Human*. Amsterdam: Elsevier, 1957.

33. Seymour, Frances I., and Alfred Koerner. "Medicolegal Aspect of Artificial Insemination." *Journal of the American Medical Association*, Volume 107, Number 19 (1936). doi:10.1001/jama.1936.02770450015004.

34. Stansfield, W. D. "The Bell Family Legacies." *Journal of Heredity*, Volume 96, Issue 1, January/February 2005. https://doi.org/10.1093/jhered/esi007.

35. Swan, Shanna, and Stacey Colino. *Count Down: How the Modern World is Threatening Sperm Counts, Altering Male and Female Reproductive Development and Imperiling the Future of the Human Race*. New York: Scribner, 2021.

36. Swanson, Kara W. "Adultery by Doctor: Artificial Insemination: 1890–1945." *Chicago-Kent Law Review* 87 (2012): 591–633. https://pdfs.semanticscholar. org/6591/55baf55869bf2251ef361186d20a5f708fb8.pdf.

37. Swanson, Kara, W. *Banking on the Body: The Market in Blood, Milk and Sperm in Modern America*. Cambridge: Harvard University Press, 2014.

38. *2020 We Are Donor Conceived Survey Report*, We Are Donor Conceived, posted September 17, 2020. https://www.wearedonorconceived. com/2020-survey-top/2020-we-are-donor-conceived-survey/.

39. Wallach, Edward, and Rudi Ansbacher. "Modern Trends: Artificial Insemination with Frozen Spermatozoa." *Fertility and Sterility*, Volume 29, Issue 4 (April 1978): 375–379.

40. Wikipedia.org.

41. Yuko, Elizabeth. "The First Artificial Insemination Was an Ethical Nightmare." Atlantic.com, January 8, 2016. https://www.theatlantic.com/health/ archive/2016/01/first-artificial-insemination/423198/.

42. Zweifel, Julianne E. "Donor conception from the viewpoint of the child: positives, negatives, and promoting the welfare of the child." *Fertility and Sterility*, Vol 104, No 3, Sept 2015. https://www.fertstert.org/action/ showPdf?pii=S0015-0282%2815%2900441-0.

Index

W

Y

About the Author

P eter J. Boni credits his disruptive childhood, a state college education from
UMass@Amherst, decorated on-the-ground
service as a US Army Special Operations Team
Leader in Vietnam (coined his "Rice Paddy
MBA"), love of his family and friendship circle, plus luck-of-the-draw DNA with making
him the person he has become today—an
author, advocate, and fun-loving grandfather
living on Cape Cod, Massachusetts.

During his accomplished business career (high-tech CEO, venture
capitalist, board chairman, award-winning entrepreneur, senior advi-
sor), Peter has applied "lessons of leadership through adversity" from
his life-altering experiences—themes found throughout his first book,
*All Hands on Deck: Navigating Your Team Through Crises, Getting Your
Organization Unstuck, and Emerging Victorious.*

In *Uprooted*, Peter intimately shares his personal odyssey and acquired
expertise to advocate for regulatory oversight of the multibillion-dollar
reproductive industry that conceives hundreds of half-siblings from a
single donor—children and adults who are unaware of the existence of
their half-siblings.

An inspiring public speaker with a storytelling, audience-participation
style, Peter enjoys an active physical regimen, entertaining and sailing with
friends and family, and traveling with his wife to, among other locales,
San Francisco and New York City to visit their growing family.